ERIKSON, ESKIMOS

& COLUMBUS

*Published in cooperation with the Center for American Places,
Santa Fe, New Mexico, and Harrisonburg, Virginia*

ERIKSON, ESKIMOS
&
COLUMBUS

Medieval European

Knowledge of America

JAMES ROBERT ENTERLINE

THE JOHNS HOPKINS UNIVERSITY PRESS • *Baltimore & London*

The Johns Hopkins University Press
2715 North Charles Street
Baltimore, Maryland 21218-4363
www.press.jhu.edu

Library of Congress Cataloging-in-Publication Data
Enterline, James Robert.
Erikson, Eskimos, and Columbus : medieval European knowledge of
America / James Robert Enterline.
p. cm.
Includes bibliographical references and index.
ISBN 0-8018-6660-X
1. America—Discovery and exploration—Norse. 2. Geography,
Medieval—Maps. 3. Early maps. 4. Historical geography—Maps.
5. Nautical charts—History. 6. Vikings—North
America—History. I. Title.
E105 .E579 2001
970.01'3—dc21
00-011396

A catalog record for this book is available from the British Library.

To my beloved wife, Esther

CONTENTS

ILLUSTRATIONS

DIRECTORY TO THE CHRONOLOGICAL SURVEY

PREFACE AND ACKNOWLEDGMENTS

UNTIL NOW HISTORIANS HAVE ALMOST UNIVERSALLY BELIEVED that Columbus's encounter with America was completely accidental. Those who ventured to think otherwise assumed that any prior knowledge he might have held about America would have sprung from Leif Erikson's contact with Vinland. New evidence presented here suggests instead that Eskimo geographical information about a wider America made its way through the Greenland Norsemen into medieval European world maps. In Europe such continental foreknowledge was not immediately correctly perceived, but it gradually drew Europe's attention westward and may have contributed to the birth of the Age of Discovery. This admittedly radical-sounding idea has grown step by step out of developments in recent decades.

On the eve of Columbus Day 1965, Yale University announced the acquisition of its now famous but controversial Vinland map, presumed at that time to depict Norse America in Canada. Yale's press release called it "the cartographic find of the century," and the map engendered two parallel disputes. The lay world saw it as an attack on a long-honored hero, Columbus. The scholarly world saw it as a completely anomalous document that thrust new problems into many branches of the study of history. Those disputes were *not* resolved when Yale announced tests in 1974 showing that the map's ink appeared to contain twentieth-century pigments. The lay world's unease regarding its hero continues regardless of that map's authenticity, for there is other evidence establishing beyond any doubt that the Norsemen did encounter North America. While to some people, forgery seemed the obvious explanation and offered resolution of the scholarly dispute, that was not the only possible conclusion from the ink tests. This author in 1977 postulated a natural scenario for the map's contemporary history that gave an innocent explanation of every known detail of the map's ink, based on modern pigment contamination, and reinstated a case for the map's credibility. Other researchers, in the 1980s, came to similar conclusions, minimizing the importance of the anachronistic ink pigments. Still others showed that the pigments might not be anachronistic at all, that they could appear in nature with the observed parameters. In 1995 Yale republished the map, the new press release stating that it "stands once again vindicated." Controversy nevertheless continues on many fronts.

Even before the ink controversy, I was working at a resolution of those disputes, entertaining a possibility that the map could preserve genuine information. However, if genuine it probably represents something other than the Canadian seaboard and something other than Leif Erikson's landfall. I will introduce evidence here to support that possibility. In answer to the scholarly problem of anomaly, I introduce evidence that the Yale Vinland Map is potentially just one member of a large group of pre-Columbian maps all apparently recording Norse contact with America or native Americans. But the Norsemen didn't *know* they knew about America. In dozens of Old World maps the Arctic coast of Eurasia shows rich though incorrect detail; Europeans had never been to Arctic Eurasia, nor had they any cartographic knowledge of it. In the past, scholars have explained away these coastlines as fantasies. It occurred to me that many of these details do correspond exactly to features on the Arctic coast of North America instead of Eurasia.

Cartographic correspondence has been discredited by its misuse in some research that reached sensational, implausible conclusions. My approach has been more rigorous, conservative, and plausible. The cartographic examples to be shown are not cryptic but very clear when viewed in the appropriate framework. Post-Columbian maps that contain apparently fanciful coastlines have been increasingly understood by careful historical analysis, and I will do the same for pre-Columbian ones. In addition to the maps, I have sought to identify numerous travelers' itineraries and geographical descriptions of pre-Columbian America, all providing mutual corroboration. While the Yale Vinland Map suffered for several decades from its uncertain origins, these other documents are of unquestioned provenance and have been catalogued in world-class libraries for centuries. However, historians seem to have overlooked or misunderstood them because, as I will suggest, the maps apparently incorporate a previously unrecognized geographical distortion that was unique to the pre-Columbian mind. That systematic distortion resulting from the misidentification of continents is analyzed herein. The thesis is elaborated in a chronological survey of geographical materials extending from the Norse Late Middle ages down through the post-Columbian Renaissance.

Did Columbus see these documents? Perhaps or perhaps not, but his conceptions of land in the west were inspired by scholars of the generations preceding his, who did see them. There is a way that ruffled feathers in the lay world may still be smoothed: I make an effort to see Columbus as more of a rationally motivated, carefully researching proto-scientist than the way his biographers have described him—strictly a luck-blessed, illogically motivated adventurer.

This work is the promised companion volume to my earlier book, *Viking America*. Neither is a prerequisite to the other, and each is written to be readable by itself. Nevertheless, issues ancillary to one are taken up in the other. (A few specific matters in the earlier book are reconsidered in the present one, as will become evident.) The prefatory comments made in *Viking America* apply here also. The polar map on the endpapers of that volume is repeated here at the beginning of the text. I refer to it throughout as the "front map" and recommend consulting it whenever the discussion enters unfamiliar regions. The more detailed local maps in Figures 2, 4, and 5 should also be noted.

This is not a history book, at least not in the strict sense of the usual deductive methodology of history from written texts. It is a prehistory book that subjects maps and documents, as artifacts, to the inductive methods of archaeology. It is based on the belief that reliable knowledge of the past can be culled from this artifactual record even if it is not spelled out in words. This work falls into place as part of the newly emerging "cognitive archaeology." I will not enter into any controversy between new and past schools but will dip into both. In examining how information about America could have reached southern Europe, I have focused on the circumstances of the various documents' creation and the lives of their creators. Known historical details will be augmented at times by scenarios grounded in a structuralist's view of human nature. Interdisciplinary support ranging from psychology to physics will be brought to bear. However, the story so constructed is not held out as proven truth. Instead it is a plausible theory to be tested against independent evidence. The day-to-day purpose of science is *not* the establishment of universal, final explanations but the proposal of testable theories that can serve as stepping stones to still better theories. Nevertheless, the more nearly universal a hypothesis, the more testable it becomes and the more potential it has for spawning still further insights. The conclusions reached here are thus a part of the sciences, subject to adjustment by new evidence or even to being overthrown—or strengthened.

This book, as well as being divided into chapters, is also structured in two parts. The second part is a wide-ranging chronological survey of documentary evidence, namely, the scores of documents supporting the theory that Norse contacts in America had European repercussions. The attention to individual documents there is necessarily limited. Part I examines in greater detail a smaller number of documents. A casual reader might, after completing Part I, wish to read just the chapter summaries and the Conclusion.

MUCH HELPFUL ASSISTANCE in the research for this book deserves to be acknowledged. Obviously, none of these acknowledgments implies any of these individuals' acceptance of the work's validity, which may or may not be expressed independently. David Woodward read the full manuscript twice in widely separated versions, giving many useful suggestions and critiques. Earlier versions were also critiqued by Benjamin Olshin and the late Vincent Cassidy. Subtopics in Eskimo archaeology received constructive attention from Robert McGhee and John MacDonald. Lunchtime conversations over the years with Thomas Goldstein gave me ongoing encouragement, and his enduring question of motivation for the discoveries was always in my mind. This research would never have been completed, let alone started, without Alice Hudson, and before her Gerard Alexander, at the New York Public Library. Also importantly, several anonymous (to the author) publisher's readers served to sharpen the arguments presented here. Last, and far from least, a highly intensive reading of a late version was given by Gregory McIntosh.

The greatest thanks to go my wife, Esther, to whom this book is dedicated. Many authors are thankful, as am I, for their wives' emotional support and encouragement, but she went further than that. She made it possible for me to work without distraction in hundreds of ways, and she devoted her professional career as a psychologist to our material support for several periods during which I pursued nothing but this research.

ERIKSON, ESKIMOS

& COLUMBUS

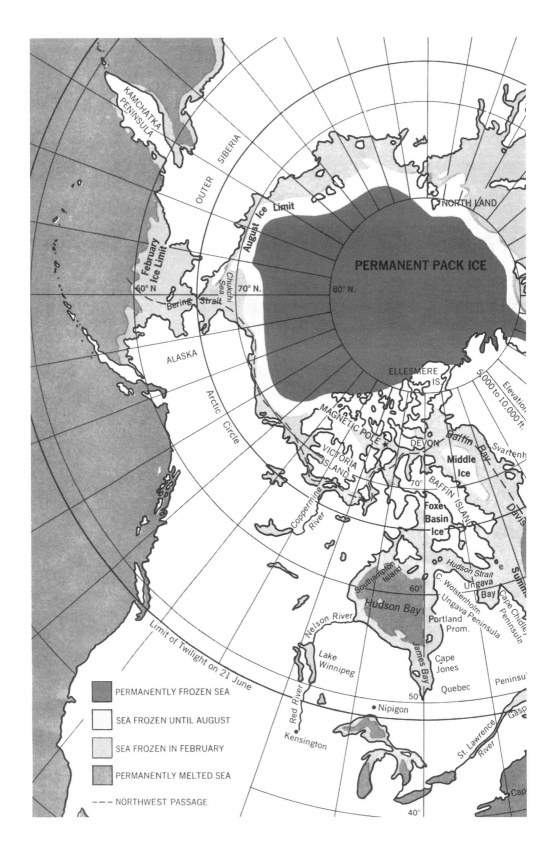

PERMANENT PACK ICE

KAMCHATKA PENINSULA

OUTER SIBERIA

August Ice Limit

February Ice Limit

60° N.

Chukchi Sea

70° N.

80° N.

Bering Strait

ALASKA

Arctic Circle

NORTH LAND

ELLESMERE IS.

MAGNETIC POLE

DEVON

VICTORIA ISLAND

Baffin Bay

Middle Ice

Svartenh

Coppermine River

70°

BAFFIN ISLAND

Davis

Foxe Basin Ice

Hudson Strait

Ungava

Summ

Southampton Island

60°

C. Wolstenholm

Bay

Cape Chidley

Elevation 5000 to 10,000 ft.

Ungava Peninsula

Cape

Peninsula

Hudson Bay

Portland Prom.

Nelson River

Lake Winnipeg

James Bay

Cape Jones

Peninsu

Limit of Twilight on 21 June

Red River

Quebec

Gasp

50°

Nipigon

St. Lawrence River

Kensington

Cap

40°

	PERMANENTLY FROZEN SEA
	SEA FROZEN UNTIL AUGUST
	SEA FROZEN IN FEBRUARY
	PERMANENTLY MELTED SEA
- - -	NORTHWEST PASSAGE

*Discovery consists of seeing what everybody has seen
and thinking what nobody has thought.*

—Albert von Szent-Gyorgy

CHAPTER I

INTRODUCTION

IN THE VIEW OF MANY WHO STUDY IT, the European discovery of America was purely a geographical event. It occurred, these people believe, when a certain lookout, weary after long sailing, cried, "Land ho!" He may have uttered those words in Spanish, Italian, Portuguese, Old Norse, Irish, or perhaps some other language. In fact, this scene arguably took place in several such tongues, leaving pedants to disagree over which was *the* discovery. Should they assign importance to priority of occurrence or to continuity of occupation? Or should they invoke some entirely different criterion to say which statue may bear the inscription "The Discoverer"?

More than either pedantry or ethnic pride is at issue, however; aside from the ideal search for Truth, there is a practical matter involved. Most American (and other) elementary schools use the discovery of America to instill certain ideals and heroic models into children's minds. The American adults who landed on the Moon gauged their accomplishment by comparison with this earlier discovery. If our schools are going to continue promoting Columbus as the most widely held cultural hero in the West, then we should understand him with some certainty.

However, our schools should probably discontinue any such heroic promotion at all. This book holds that the very concept of a European "discovery" of America is clouded by a reciprocal concept of divulgence by native Americans encountering the Norsemen. Ethnic priorities become secondary to an intellectual development that ultimately involved all the peoples engaged in a protracted uncovering.

Serious thinkers already recognize that the discovery of America was a much more complicated happening than merely a landfall. Indeed, it took some decades after 1492, before most geographers realized that there was an

America distinct from Asia. Until that time Europeans, with infrequent exceptions, presumed that any mainland on the other side of the ocean from Europe must be Asia. The many explorers who were active in those decades after 1492 are the ones who really deserve to be called the discoverers of America in the sense of discovery pronounced by Szent-Gyorgy in the epigraph to this book. Columbus, island hopping in the Caribbean, never set foot on new world mainland until at least 1498, probably 1502, and that only briefly and initially unknowingly. Even when he came to "know" that a new world was involved, his state of knowledge was vague, theoretical, and basically a misunderstanding that sprang from the traditional presumption of "antipodes."

These statements may already be unsettling to some who have been indoctrinated with the heroic concept of Columbus. Those people may hear such statements as merely iconoclastic. Or they may come to his defense with the thought, "But such explorers would never have gone there at all if it were not for Columbus." This thought *may* be so, but one must observe very closely whether it is summoned in defense of Columbus the man or Columbus the creed.

Nobody can ever deny that Columbus the man was important in the discovery of America. Columbus's demonstration of the possibility of crossing the middle latitudes of the "Ocean Sea" remains a bold, epoch-making undertaking. But we also want to evaluate his situation rationally. First one must make a wholehearted attempt to see him as an ordinary human being, thinking and reasoning in familiar ways and driven by familiar, fathomable desires. If he does not end up seeming larger than life, that is no criticism of the man but a reevaluation of what we have done with our minds in the classroom.

Leif Erikson, too, was a discoverer of America. There is no longer any controversy among scholars about that. In his case the bias in teaching had been in the opposite direction. The saga of Leif's voyage to Vinland ca. A.D. 1000 had always been introduced with an apologia, almost clandestinely, as a rumor that very well might not be true. Teachers felt it should nevertheless be mentioned for the sake of fairness and completeness. There are discovered ruins, now, to prove it—the Norsemen were in America.[1] Furthermore, their culture was not the barbaric way of life from earlier centuries depicted in popular media; its technology was similar to that of many small towns and rural areas in medieval Europe.

Nevertheless, Leif's discovery was somehow premature; for some reason it led to no obvious repercussions in Europe. Columbus's discovery was somehow more timely. What does that word *timely* mean? A standard answer is often couched in European political and economic developments of the late fifteenth century. Then-recent expeditions out into the ocean looking for islands are cited as important precursors. While such factors obviously were involved, they do not tell the whole story either. The Columbian discovery of America was not only an event in the expansion of Europe but also an event in intellectual history, the history of ideas, indeed, the history of science. That is the area in which I address the timeliness of the Columbian discovery.

The above-stated thought that later explorers would never have gone to America had Columbus not done so may be wrong. The *idea* of sailing westward for Asia was not unique to Columbus. Scholars know that some of his contemporaries and predecessors entertained the same idea and that a few of them actively pursued financial backing for a trip. The established attitude that Columbus had to battle was *not* concerning the sphericity of the earth. Contrary to what our schools have taught children, those with a practical concern in the matter had known ever since Greek times that the earth had to be round.[2] But they also knew something about its size. Many inferred from astronomical measurements that the distance one would have to sail westward from Europe to reach Asia was probably 12,000 miles or more. They knew that a ship would have to carry a year's worth of provisions to attempt such a voyage safely. No crew or captain was willing to sail that long out of sight of land exposed to even the predictable dangers, and therefore the voyage was considered impossible. This was the establishment's argument in Spain against Columbus, and they were right.

Yet somehow, during the fifteenth century an attitude began to proliferate holding that the earth was much smaller than indicated by astronomical measurements. Columbus believed Asian shores to be only one-quarter the above distance away, and similar arguments by various near contemporaries gave him courage. Even without Columbus, America would very likely have been discovered within a few years or decades. The project was "in the air."

The question of timeliness now takes on a new interest. What does "in the air" mean? Structuralist psychologists, led by Jean Piaget's examples, have begun to understand scientifically the concept of a "body of knowledge" and the modes of transition of knowledge from one state to another.[3] One basic tenet is that people cannot assimilate or appreciate a new idea until they can connect it logically to some contemporary established knowledge— something already known.[4] Thereby they accommodate to the new idea. The underlying question now seems even more confounding. Why did the completely incorrect idea—that the shores of Asia lay only a few thousand miles across the ocean—gain credence *at this particular time?* To what canonical knowledge or gestalt or idea structure was it attached?

When the seekers after Asia in the West tried to enlist financial backing for a voyage, they needed evidence to support their theory. One argument Columbus and his supporters used was that classical and medieval writings, including those of Aristotle, Ptolemy, and Marco Polo, contained geographical references suggesting the near proximity of Asia. It seems possible however, as Samuel Eliot Morison thought, that such authorities were merely summoned up as afterthoughts[5] employed in a wide-ranging attempt to buttress an argument they already believed in.* Something unique to their own era might better explain the timeliness of the ideas of Columbus and like thinkers. Something stronger than (erroneous) ancient authorities perhaps made them invest years at royal courts and risk their lives at sea defending an argument about the distance to Asia that many contemporary theoreticians rightly scoffed at.

Yet neither they nor later students of their ideas have been able quite satisfactorily to point out what carried them to such unjustified certainty. Motivation for belief (above all, profit) there was aplenty, but for certainty, none. Some have concluded that Columbus was probably mad.[6] Many writers concentrating on Columbus's motivation have completely overlooked the motivation of the others who proposed such voyages. Were they all mad? The problem has been so frustrating that some writers have turned to mysticism for an explanation.[7] Jean Piaget, the founder of structural psychology and scientific epistemology, would have allowed no such easy escape. He stated flatly, "The great man who seems to launch new movements is but a point of intersection or of synthesis, of ideas that were elaborated by continuous cooperation. Even if he is opposed to current opinions he responds to underlying needs which he himself has not created."[8] While structuralism has been applied in certain inappropriate fields, and thereby discredited, current research is confirming and extending Piaget's concept.[9]

In *Viking America* I suggested (as others had before me) one stimulus that could have given rise to belief in the nearness of Asia: vague and misunderstood but actual information concerning America.[10] The present book

---

*Even Ptolemy's *Geographia,* which was first translated from Greek into Latin around 1406, could have been available to Western scholars long before that if they had any interest in the subject. (See p. 12 of Milton V. Anastos, "Plethon, Strabo and Columbus," *Annuaire de l'Institut de Philologie et d'Histoire Oriental et Slaves de l'Université de Bruxelles,* vol. 12 [Brussels, 1953], 1–18). Furthermore, if the *Geographia* were as exclusively important as some writers would have it be, its influence should have been felt two generations before Columbus.

shows evidence that residents of the Norse colonies on Greenland transmitted such information, in the form of maps, travelers' descriptions, and living natives, to Europe during the century or so preceding Columbus. And in the recentness of those particular Norse contacts lies a possible explanation. A fourteenth-century resumption of westward exploration after centuries of neglect could explain the timeliness of the southern European ideas of the proximity of Asia in the West. While Leif Erikson's eleventh-century contacts in Vinland ultimately resulted in abandonment of Vinland, those of the early Renaissance Greenlanders might have led toward the European "Age of Discovery." And that resumption of the Norse exterior orientation, as we shall see, was itself a result of new contact with the native American people we call the Eskimos.* The discovery of America would be an involved joint accomplishment not solely attributable to any individual hero or ethnic group.

Numerous writers who surmised that Columbus had some kind of Norse information have been dismissed for giving no proof. The reader is specifically put on notice that I make no claims of final proof here regarding Columbus personally. I leave it to others to draw conclusions in that regard, perhaps after further research. But the evidence that such information was *available* in southern Europe in his time will be substantially increased, and we will see some tentative evidence that Columbus had access to it. Historian Samuel Eliot Morison, in his biography of Columbus, concluded that he had an illogical mind. While asserting that Columbus's theory was dubious, Morison said, "One can well imagine him explaining it, his eyes sparkling and his ruddy complexion flaming."[11] There were undoubtedly many factors contributing to his motivation besides rational ones.[12] Indeed, current writers on the subject have highlighted various aspects of that motivation that are nowadays considered politically incorrect. But we will see that a rational component to his motivation also becomes a possibility.

However, I do not think, as some media headlines suggested when *Viking America* was published, that Columbus secretly followed Viking maps. Nor do I claim that Columbus necessarily engaged in any kind of major deception (although others have shown evidence for such theories).[13] The process of any information transfer from the Norse was likely much more subtle than conscious deception. Structuralist theoreticians acknowledge the possibility

*Modern descendants prefer the term *Inuit,* or in Greenland *Kalaalit,* but we will be dealing with older sources and will retain the term *Eskimo.* For the same reason we will use the term *Greenlander* to refer to the medieval Norsemen rather than the modern *Inuit.*

of unconscious response to vague influences.[14] The "new historicism" researchers find many traces of unconscious behavior during the encounter with the New World.[15] Louis DeVorsey has shown how in post–Columbian times native American information sources "silently" entered into European maps.[16] G. Malcolm Lewis showed that similar information led to systematically misunderstood representations on European maps of America which motivated explorations even if they were ultimately dismissed as myths.[17] Surely, that kind of vague influence could have affected the reactions of Columbus and his predecessors and contemporaries to any Norse-transmitted foreknowledge about America.

If the importance of the Norsemen and the Eskimos in European history is elevated by this book, it is not meant to be at the expense of Columbus or his bravery. Columbus also will be elevated, for he will become more understandable as a person. The man becomes an intelligent, rational participant in one of the greatest human accomplishments to date: the development of the scientific method. It is an auxiliary purpose of this book to examine evidence in that direction.

It will be obvious that this book contains radical ideas and uses radical procedures. In order to make a positive approach under these conditions, it will be useful to review (and preview) some fundamental considerations: the cultural milieu of the interaction between the Greenland Norsemen and the Eskimos, within both the European culture and the Eskimo culture; and the European intellectual milieu in which the Norse Greenland colonies matured, particularly with regard to geographical thought. This will introduce how maps looking like American lands could show up on medieval and Renaissance maps of Eurasia.* Such an idea, however, will immediately raise doubts in scholars who are skeptical about interpreting old maps by their shapes. This leads finally in this chapter to a section on method, to assure that we are proceeding with due caution.

I HAVE EXAMINED THE PHYSICAL DETAILS of the Norse contact with America, in *Viking America*.[18] I suggested that there was a renewed contact after centuries of hiatus during Early Renaissance times, involving Eskimo contacts.† Some evidence suggests Norsemen made physical contacts westward

---

*The Renaissance is defined officially to have begun during the fourteenth century, but the separate attitudes characterizing the Middle Ages and the Renaissance coexisted for centuries. We may consider the Middle Ages to have ended with Columbus.

†I will not maintain here, as I did in Viking America, that the Norsemen actually assimilated into the Eskimos. That will be left as an open question, although it seems increasingly unlikely. Fur-

into the Arctic Archipelago (see north of Canada on the front map) and perhaps into North America itself. But the cultural contacts will be our main interest here. The Norsemen lived in fully civilized communities that kept in touch with the European world. Their so-called Eastern Settlement was near the southern tip of Greenland's west coast, and the Western Settlement was just a little farther up the west coast. Their summer hunting area, Nordrsetur, was above the Arctic Circle on the west coast. When fourteenth-century climatic changes forced a decline in their agricultural and herding settlements on Greenland, the Norsemen's lifestyle became more nomadic and they lived increasingly by hunting. Written sources show that they soon encountered the Eskimos in Nordrsetur. (A brief eleventh-century encounter in Vinland seems not to have been sustained—at least no evidence suggests it was. Southern Greenland itself was not occupied by living Eskimos when the Norsemen first arrived.)

In the nineteenth century it was fashionable to believe that hostile interactions between the newly arriving Eskimos and the established Norsemen caused the eventual end of the Norsemen in Greenland and that there was little interaction otherwise. In the twentieth century, Gwyn Jones continued this idea, in concurrence with many European archaeologists. But Fridtjof Nansen and Helge Ingstad both rejected a fundamentally hostile interaction, and they have many followers in North America.[19] Nansen and Ingstad both concluded that, even though individual incidents of hostility certainly did occur, the overall interaction was generally neutral and sometimes positive. Danish archaeologists now concur in rejecting the extermination hypothesis.[20] The thirteenth-century Eskimos that the Norsemen encountered were engaged in the culmination of a migration across the Arctic from Alaska to Greenland. These Eskimos, members of the Thule culture, might have set examples for the Norsemen on how to live in the high Arctic winters and where to find hunting grounds.* In their turn, the Norsemen have been thought to have set examples for the Thules in many technologies (like baleen saws and coopered tubs) that the Thules incorporated into their daily life. The result was construed to be the emergence of a new Eskimo culture, the Inugsuk, recognized by the Scandinavian archaeologist Therkel Mathiassen

---

thermore, while the contacts were widespread, they were not necessarily frequent and probably involved an element of accident.

*The word *culture* in this usage is a modern anthropological construct used to differentiate one group with a common lifestyle history from another, for example the earlier Dorset culture Eskimos. Each has a distinct spectrum of tools and artifacts in the archaeological record. The name of the culture comes from the location where it was first discovered by archaeologists.

and amplified by Erik Holtved and Junius Bird.[21] Some archaeologists still consider any such cultural interaction between the Norse and Eskimos to be controversial.[22] Nevertheless, the Canadian archaeologist Robert McGhee, who is considered to be the "dean of Canadian Arctic archaeology," sees evidence of positive cultural interaction. He believes it is likely that Eskimos and Norsemen had wide-ranging trade contacts and coexisted, sometimes amicably, for several generations in southern Greenland, leaving at least one shared historical tale in the folk record of each.[23] Norwegian linguist Knut Bergsland has cited Norse loan words for domestic plants and animals in the Greenland Eskimo dialect.[24] Danish National Museum archaeologist Jette Arneborg has given theoretical arguments that we should also expect to find cultural influences in the opposite direction, from the Eskimos to the Norsemen.[25] While no one has yet been able to demonstrate any in the archaeological record, we find some in this book.

I suggest that the Thule Eskimos occasionally shared their cumulative geographical knowledge of America, gained during the recent migration, with the Norsemen. There is no more pressing question about such a newly met people than, "Where on Earth did you come from?" Neither party has to be bilingual to pose the question or give an answer, as many explorers have demonstrated. The Eskimos' geographical knowledge included the technique of drawing primitive maps—cartograms (surprisingly not part of the Norse practice). The idea of the map, and skill in its realization, have been part of many nomadic, so-called primitive, cultures because they assist survival. The medieval Thules were a primarily coastal people living from sea mammal hunting, and they were acutely aware of every turn of the coastline. The pioneer anthropologist Franz Boas visited their descendants in the early 1880s, before they had contact with modern Europeans. About them he said, "If a man intends to visit a country little known to him, he has a map drawn in the snow by someone well acquainted there, and these maps are so good that every point can be recognized."[26] Figure 1 shows an Eskimo map of Southampton Island (at the top of Hudson Bay—see front map) that Therkel Mathiassen collected on the Fifth Thule Expedition. Its only major shortcoming is local variation of scale. Note a doubling of scale in the eastern peninsula. A similar doubling in the southeast coastline along Fisher Strait has caused a rotation of orientation of other features. More recently, Spink and Moodie thoroughly surveyed the question of primitive Eskimo maps and concluded that they were part of the indigenous culture. Certainly no one denies that the relatively sophisticated geographical awareness that goes into creating a map was part of the Eskimos' culture.[27] Part of the difficulty in

a

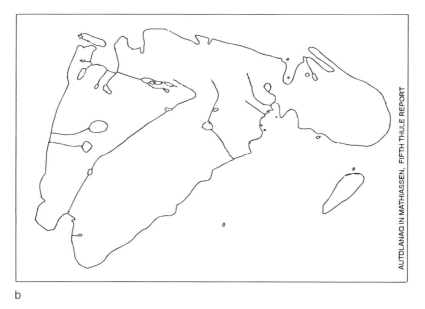

AUTDLANAQ IN MATHIASSEN, FIFTH THULE REPORT

b

Figure 1. Modern aerial survey map of Southampton Island (*a*) compared with "primitive" Inuit Eskimo map (*b*)

recreating a properly scaled map undoubtedly arises because much of the information was memorized in a linear itinerary-like fashion.[28] It is known that Greenland Eskimos did carve maps on wood, which were apparently linearly organized.[29] The suggestion that Eskimos only learned mapmaking from post-Columbian contacts with Europeans is supported by no evidence and is motivated only by chauvinism about mapmaking as an advanced, "civilized" technology. Agriculture-based societies have always looked down on the hunter-gatherer mind as "primitive," yet in reality the latter are just as capable.[30]

Most of the maps we will examine more likely originated from Eskimo information than from Norse. Furthermore, they probably would have originated from contacts more or less contemporaneous with the appearance in Europe of analogues of each map. In fact, many might have been the result of European cartographers directly interviewing Eskimos brought to Europe rather than transfer on paper. This is particularly true of some early maps resembling the Alaska area, where the Thule culture originated.* Eventually the maps showed features incompatible with an Eskimo origin, such as evidence of surveying involving latitudes. This would indicate that the Greenland Norsemen, who kept intellectual contact with Europeans, had also learned some notions of cartography from the Eskimos. Indeed, as the Norsemen eventually became nomadic hunters themselves when their climate deteriorated, that cartographic skill became of vital importance to them also. Such later maps may well have originated directly on permanent media.

The history, nature, even the very existence of any such cartography on American shores in the Middle Ages must remain conjectural. Even the word *cartography* may be inappropriate, for much of the geographical information may have been carried in the head, as it was by Boas's Eskimos, and drawn on ephemeral media, if at all, or described verbally.[31] More recently Edmund Carpenter described his observation of Eskimo hunters: "When a man travels by sled into unfamiliar country, he continually looks back to see how the country will appear on his return. These brief glimpses, vividly recorded and faithfully remembered, are enough so that he can find his way back with ease. An elderly hunter may efficiently guide a party through an area he has not seen since his youth, and then but once." Exactly this kind of performance was also witnessed by anthropologist Hugh Brody. This is not a racist attribution of special powers but suggests a cultivation of perceptions that the environment requires for survival. Everyone has heard the statement that the Eskimo language contains scores of special terms dealing with snow conditions.[32] The language also contains a large special vocabulary devoted to orienteering and geographical terms. Robert Rundstrom has shown by struc-

---

*This is a retraction of a suggestion made in *Viking America* that the Norsemen themselves attained Alaska. The Eskimo origin interpretation is more plausible and has much evidence to support it.

tural analysis of the culture that the cartographic notions were just another aspect of a characteristic Eskimo mimicry of the environment, a mimicry also noted by Carpenter.[33] The notions were just as likely to have been present in the Middle Ages during the migration as more recently. Indeed, such cartographic notions were probably a major contributor to the remarkable speed of that migration.

In medieval Europe, however, the idea of the scaled map detailing local features was seldom encountered in everyday life. Geometrical and geographical information was transmitted in elaborate verbal descriptions supplemented at most by pictorial diagrams.[34] Thus, even though I will continue to use the words *map* and *cartography,* the reader must bear in mind that these will indicate broader concepts of information transfer as well as paper drawings. The detailed history of any such transfer obviously cannot be shown. Indeed, if there were any hard-copy prototype source maps that we might otherwise hope to have found in the historical record, they were probably immediately discarded. Such has been the fate of the majority of medieval maps. Once incorporated into world maps by the few cartographers who could accommodate to them, any localized cartographic materials had no further purpose. P. D. A. Harvey stresses that before the popularization of Ptolemy's chorography,* local maps, as opposed to world maps, were practically unheard of in medieval Europe.

So, there is no hard direct evidence—and we should no longer expect to find any—of any cartographic influence from the Eskimos to the Norsemen and Europeans. Therefore, we will be unable to provide a deterministic or deductive history in this matter. But this does not necessarily limit us purely to speculation. Nor does it necessarily require us to remain permanently in the dark. We can still make some progress by proposing a formal *hypothesis* that such transfer did in fact occur, and then examining the supporting evidence. There need not have been extensive contact, just enough to stir intellectual curiosity each time it occurred. Such a postulate is obviously not only plausible but perhaps likely, even though it remains to be sustained. In archaeology, such hypotheses are frequently invoked to make cultural sense out of a large collection of objects and observations. In our case, the observations are not field archaeological but archival archaeological, the world maps and texts we shall look at. This postulate, we will see, makes it possible to weave together a large number of odd, unexplained and previously unre-

---

*Ptolemy's word for the mapping of features restricted to a locality sufficiently small so that global projection considerations are unnecessary.

lated observations into a unified fabric, a hypothetico-deductive model. This is the essence of philosophy, even if the fabric remains to be tested. Passing the test converts it to science.[35]

If this fabric withstands the upcoming test, it will suggest that there was *no* specific "first" European discovery of America as we think of it, not even by the Norse; instead, there was a gradual Eskimo divulgence to Europe of continental land existing just west of Greenland. Only much later would anyone conceive that this was *not* Asia.

MEDIEVAL WORLD MAPS IN EUROPE were in the form of experiential, topological descriptions (called "cartograms" by some writers) rather than rigorous grid-referenced data—until Ptolemaic principles reemerged. Early European maps have a bad popular reputation that can be blamed on a special aspect of the human brain. Our visual apparatus has evolved to put great emphasis on recognition of the finely detailed features of familiar human faces that make them familiar. That same equipment detects minor variations in the shapes of maps. It is true that early cartograms departed from what is familiar now, but most at least recognizably suggested reality, certainly in the fourteenth and fifteenth centuries. At least they were based on information, right or wrong.

The cartographic depiction of Scandinavia as practiced in southern Europe has a very curious history. Scandinavia had no native tradition of making maps, and for many centuries European maps depicted it quite minimally. (Nevertheless, Scandinavian scholars were quite aware of and kept up with southern European cartographical and cosmographical developments.)[36] Then in the fourteenth and fifteenth centuries many bizarre shapes suddenly appeared in the northwest corner of European maps, as we shall see.[37] But though these shapes had much variability in their unreality, they showed little or no evolutionary development toward recognizability during these two centuries. (A few maps could be construed to have some resemblance to Scandinavia, but few indeed and showing no effect on any evolution.) Meanwhile, other parts of the world map evolved recognizably. This atypical circumstance concerning Scandinavia terminated abruptly early in the sixteenth century with the appearance of known Scandinavian cartographers. I will uncover at length a rational explanation of this development as a natural byproduct of our hypothesis of Eskimo information divulgence.

The European world gradually lost contact with the Norse Greenlanders, and there is no record of any *official* communications after the early 1400s. Nevertheless there are many suggestions that Norsemen continued living there and that Europeans continued going there for another century. Late-fifteenth-century European fashions have been found in Greenland burials. Vague bits of information about the lands to the west, as we shall see, might have gotten through to Europe by various unofficial, unrecorded channels, both before and after 1400. A similar thing happened with information from unrecorded New World explorations after 1500.[38] What would have happened to such information once it arrived in Europe? Part of the job of many cartographers, at least in the *late* Middle Ages, was to incorporate new data when creating a new map. Once in Europe any American data would have given rise to a classical situation in Piagetian epistemology: how were any such odd maps, descriptions, and geographical scraps to be understood in Europe? They could not be connected logically to any established knowledge structure. Mostly, I suggest, they would have been completely ignored and discarded rather than divulging the new continent.

How can something stare you in the face and yet be unseen and unmentioned? It happens all the time. According to the structuralists, an organism simply cannot respond to a stimulus unless the stimulus has some rudimentary meaning or is somehow known to the organism.[39] Infants exhibit this phenomenon when first presented with one object disappearing inside or behind another and reappearing: they actively refuse to look at the anomalous situation that seems to violate the permanence of objects.[40] Later, in the "peek-a-boo" stage, they are able to accommodate to this anomaly by considering it fascinating, humorous, and entertaining. Similarly, most European intellectuals would have simply rejected as preposterous evidence of mainland so close in the west.

Nevertheless, a few geographers might have found enough hooks in the established state of knowledge to be able to assimilate the new information. One element of the established state of knowledge was, as previously stated, that no maps or information about the geography of Scandinavia existed. Authoritative European cartographic knowledge of Scandinavia was a vacuum. Meanwhile, consider the hypothetical occasional maps of or information about American lands transmitted by Greenlanders that might have come into the hands or ears of European cartographers. These maps—or the individuals with the information—would naturally have reached Europe by way of Scandinavia, so some cartographers seem to have jumped to the conclusion that the cartographic information must therefore relate to the geography of Scandinavia! It turns out that the unrealistic outlines of Scandinavia were fairly good likenesses of various *American* lands. The northwest corner of the map of the Old World appears, on various occasions, to have assimilated outlines with origins ranging separately from Labrador all the way along the Northwest Passage to Alaska, mirroring the route of the Thule migration, under labels of Scandinavia. I will refer to such a mental telescoping hereafter as the "provenance paradigm," indicating a presumption that any novel geographical information may be associated with the locale of its immediate provenance.

Thomas Kuhn introduced a definition of *paradigm* as a framework of assumptions that guide the vision of a field of study—a canonical way of handling problems.[41] He showed that within a given field, practitioners tend to adhere unquestioningly to the established paradigms for long periods of time, giving them up only when a revolution in the field establishes a new paradigm. My provenance paradigm may seem rather trivial compared to some Kuhn studied, but it meets all his criteria in our restricted field. European cartographers thought, "If the data comes from Scandinavia, it must be data about something in or near Scandinavia." Why think any harder, especially if it leads to the preposterous?

During the sixteenth century, learned Europe underwent a revolution in another larger paradigm, the paradigm within which global geographical thought took place. This new global geographical paradigm, which we practice still today and which did not become firmly established until many decades after Columbus, teaches that the globe actually contains a multiplicity of distinct, unconnected, continents in opposite hemispheres. Before the sixteenth century, beginning in antiquity (with exceptions to be noted below), the established paradigm held that all the Earth's mainland, excluding islands, existed in a single interconnected continental cap, *terra firma*. The very meaning of the word *continent* originally was "held together." That world continent is known to us even today as the Old World, comprising interconnected Europe, Asia, and Africa. It was known throughout the Middle Ages in Latin as *Orbis Terrarum;* in Old Norse, *Heimskringla*, the "home circle." The presumed circular outer perimeter was geometrically imposed by theory as follows. Supposedly the globe was made up of a land sphere and a water sphere, mutually interpenetrating in the same global space. (Water soaks into land and land sinks under water. There was obviously no concept of the core.) Slightly off-centered from one another during Creation, the land world rose above the ocean surface but was still attainable by anchor chains if one was not too far at sea.[42] In such a globe the boundary between dry land and ocean is necessarily circular if the land sphere is comparatively smooth, allowing for minor deviations such as bays and peninsulas. And, most importantly, in such a world there can be only one ocean and one *terra*. In the Bible, God said, "Let the waters under the heaven be gathered together unto one place, and let the dry land appear" (Gen. 1:9). I shall refer to this outlook henceforth as the "one-ocean paradigm," reflecting the original sense of *the* one global ocean, Homer's all-encircling Oceanus become global.

Alternatives to this paradigm, which arose from time to time throughout history, speculated about quadripartite worlds, multiple oceans, and antipodes.[43] But there were pressures on most scholars to remain within the one-ocean paradigm. Such pressures have been inherent in the established paradigms of all fields of scientific study and are probably related to fundamental psychological processes.[44] But this fixation left its practitioners open to being misled by any potential contacts with America. According to the one-ocean paradigm, an American mainland would have had to be seen as Asian. The most famous alleged practitioners of such a one-ocean assimilation of America were, of course, Christopher Columbus and his followers, until the existence of the New World was recognized.*

We will see evidence that Columbus was not the first to be misled in this way. Hypothetically, with the expansion of the Greenlanders' knowledge, direct or indirect, of the continental Arctic coast and perhaps down the Canadian east coast, the continental nature of the land would become apparent. While European maps had for some time been assimilating *local* American places as part of Scandinavia under the provenance paradigm, they could not assimilate this continental aspect using that paradigm. So, the cartographers hypothetically concluded, the preposterous was true: it must be eastern Asia. This was the only remaining alternative to postulation of a new continent, an even more radical concept. (Northern and eastern Asia were then as much a cartographic vacuum as Scandinavia had been.) From the early fifteenth century onward, southern European scholars increasingly seemed to try to accommodate their concepts by addressing the question, "How far westward would one have to sail to reach the continental land of Asia?" During that century, as we shall see, many European world maps displayed outlines of northeastern Asia that resembled the northeast corner of America.

The one-ocean paradigm, it must be realized, is a total global concept. North America would be seen as Eurasia and even, perhaps, South America as Africa (which is southerly from Eurasia and attached to it by a narrow isth-

---

*While Columbus never personally set foot on North America, Caribbean Indian stories about the continent led him to believe he was near Asia. He sailed briefly along Central America without thinking it was continental. After his later, brief encounter with South America, he thought of it as the theoretical Antipodes, not the ocean-cleaving New World we know.

mus).* The paradigm can invoke a view of eastern America as eastern Asia, but theoretically it could also invoke a view of western America as western Europe. Thus, any data concerning the north*west* part of America (Alaska) would have had to appear on Old World maps in northwestern Europe (Scandinavia). Since the result of this is graphically equivalent to the result of the provenance paradigm, an ambiguity about the means of adaptation would arise when Alaskan land appears (which it indeed does). When a map displays such a configuration, I sidestep the ambiguity, not attributing it to either paradigm in particular, just both collectively.

So, we are going to see evidence that there were two counterbalancing forces in effect for several centuries. On one hand, there were information sources creating a possibility to divulge the existence of the American continent. On the other, there were paradigms in operation that served to hide any such divulgence. I refer to the provenance paradigm and the one-ocean paradigm collectively as the "divulgence-hiding paradigms." In *Viking America* I used different terminology for these, labeling the one-ocean paradigm the "Grand Misunderstanding" and the provenance paradigm the "Smaller Misunderstanding." I will continue to use the term *misunderstanding* when referring to specific instances and occurrences, but the paradigmatic concept is much more useful for understanding the mental processes.

I attempt to demonstrate, for each European map studied, an American prototype area that the paradigms transformed into the resultant map. This approach involves minute attention to the details of coastal and interior features that are preserved during the transformation. Indeed, the necessary unhinging of one's visual and mental expectations may sometimes seem disconcerting.

THE ESKIMOS WERE THE SUBJECT OF A NONCARTOGRAPHIC MANIFESTATION of the divulgence-hiding paradigms in Europe. When stories about the Eskimos reached southern Europe, they apparently were misinterpreted as applying to legendary peoples in northern Eurasia. The historical myth about such people of the North is closely associated with the myth of the Arimphians and the Riphean Mountains, the great stone girdle which the Russ people believed encircled the whole earth. While legend placed the Arimphians south of the Riphean Mountains, a people called the Hyperboreans

---

*Some versions of Ptolemaic maps closed off the Indian Ocean in the east, which would have prevented conceiving Africa as attainable from eastern Asia. However, this idea was far from universal.

were placed just north of the Ripheans. Among others, the medieval historian Adam of Bremen (ca. 1073–76) wrote of the most northerly people in the world, the Hyperboreans. The name and myth of the Hyperboreans probably originated in the ancient Greek worship of Apollo.[45] As the myth evolved and the Hyperboreans became displaced northward in later centuries, the description of the Hyperboreans became more and more potentially applicable to the American Eskimos. The Hyperboreans (and Arimphians) were always emphatically described as a "happy people."[46] Such a description is eminently applicable to the Eskimos. The first and most vivid description that all early modern writers on Eskimo life gave was of the Eskimo happiness and amiability. When Eskimos were eventually brought by Norsemen into contact with Europeans and interviewed by learned men, the temptation would have been strong to identify them as Hyperboreans.

Indeed, the standard medieval way of incorporating knowledge of newly encountered peoples was to try to identify them with something already in the traditional record. This is the approach described by structuralist theory, looking for hooks of attachment. In this way, the Mongols later became associated with the legends of Gog and Magog, the enclosed nations it was feared world start the Armageddon. Another remarkable characteristic of the legendary Hyperboreans was that when they attained old age, they threw themselves into the sea from a cliff. This strongly resembles the Eskimo practice of self-sacrifice in old age.[47] Numerous modern explorers habitually used the term *Hyperboreans* to refer to the Eskimos they met, and to this day the languages of the Arctic peoples are classed as Hyperborean languages. A central concept of the classical Hyperboreans was that they were worshipers of the Sun. While the Eskimos' religion was primarily animistic and shamanistic,[48] they did have many superstitions and myths regarding the Sun that would have fed the mistaken identification. They engaged in elaborate festivals of the winter solstice and believed in many practices intended to encourage the quickened return of the Sun from the winter darkness.[49]

Columbus and his successors, operating within the one-ocean paradigm, identified native Americans as "Indians." Those who were brought back to Europe learned European languages and related much about their homeland. The Norsemen are also known to have brought Eskimos into contact with European civilization. Karlsefni brought back "Skraeling" natives from Vinland. Circumstances suggest that these were Eskimos rather than Indians. Are Frode's early-twelfth-century *Islendingabók* described a Norse archaeological survey of Eskimo remains (presumably Dorset) in Greenland: "in both the east and west of the land they found dwellings and fragments of boats and implements, from which they perceived that the people who lived there were of the same kind as those of Vinland, whom the Norse Greenlanders call Skraelings." Fourteenth- and fifteenth-century contacts were clearly with Eskimos, identified by their kayaks, their latitude, and other distinctive details. In 1385 world traveler Björn Einarsson took two young Eskimos into his household while in Greenland. If any Eskimos were ever taken physically to Europe, they would likely have been perceived as Hypereboreans.

An area of Alaska occupied by the western branch of the Thule Eskimos and their immediate predecessors, the Birnirk culture, is shown in Figure 2.[50] It contains a feature concerning the Sun that the legend of the Hyperboreans emphasized, the transition from familiar days of daylight and darkness to the phenomenon of midnight sun and extended days (and alternately nights). That transition, of course, occurs at the Arctic Circle. Until recently most archaeologists and anthropologists thought little about the Eskimos' astronomical concepts, but new investigations by John MacDonald show that the Eskimos were astronomically quite aware.[51] In particular, they were aware of the transition at the Arctic Circle. No sophisticated global concepts are needed simply to observe the difference in the Sun's behavior on either side of the transition. All that is needed is an awareness of the importance of the Sun in daily activities, which they certainly had. In winter above the Arctic Circle during the "Great Darkness," they fought a disturbed circadian sleep cycle by keeping time with the stars; the return of the spring sun was accurately anticipated and measured in quantitative stages.[52] Right at the Arctic Circle on the day of the winter solstice, the Sun is hardly seen at all; but the next day it resumes its usual appearances, providing a nontechnical means of specifying this location. Several Eskimo myths address exactly this phenomenon.[53] Being a primarily coastal people without sophisticated global concepts, the original Thules would not likely have had much experience or concern with the fact that this transition occurs all along the Arctic Circle through the continental interior. To them, the transition would have been associated with the unique area along the Alaskan coast near Kotzebue Sound and Seward Peninsula. Later in this book we will see cartographic traces of this Alaskan area right at the Arctic Circle, but placed in Scandinavia, its appropriate place under the divulgence-hiding paradigms.

THAT PEOPLE SEE THE COASTLINES OF ONE PLACE in a map that purports to be of another place has always posed problems for historical cartographers, who accept as a truism that appearances can be deceiving. The phenomenon has been referred to as "wishful seeing." In other fields incorporating large

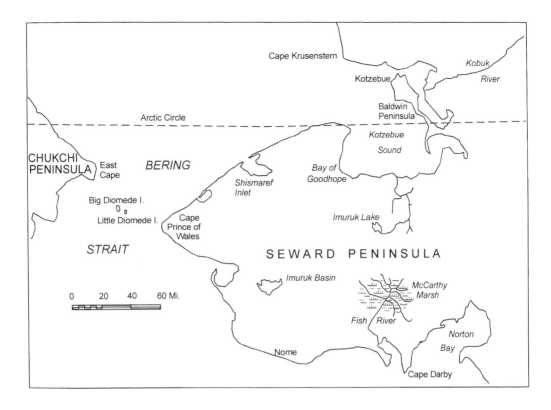

Figure 2. Bering Strait, Seward Peninsula, Kotzebue Sound area of Alaska

collections of data, the process of searching for novel connections has been formalized as "data mining" or even "data dredging."[54] One famous map by Piri Reis contains shapes that have led Charles Hapgood to unlikely but spectacular conclusions about Antarctica that the scholarly world rejects.[55] Another group of maps has been taken by Pablo Gallez to represent the Strait of Magellan even before Magellan went there.[56] Less spectacular and more plausible explanations of these maps have been given.[57] There are many other cases of map interpretations leading researchers astray. An illustrative students' game in the subject is to turn a recent map of some locality upside down and then to try to rationalize it as a medieval map of some other place.[58] Even respected scholars in the field have come up with multiple contradictory interpretations of the same unidentified coastline. In the 1960s mathematician Benoit Mandelbrot formalized the idea that coastlines in nature exhibit fractal self-similarity: "Seacoast shapes are examples of highly involved curves such that each of their portions can—in a statistical sense—be considered a reduced-scale image of the whole."[59] In particular he observes, "It is hard to tell small and big islands apart, unless one either recognizes them or can read the scales. . . . many islands look very much like distorted continents."[60] The point is, one must exercise the greatest care when trying to interpret maps the way I propose we do.

How does one exercise such care? Exactly what should the methodology be? In the 1970s one answer to that question began to emerge as a movement within the community of historical cartographers reacting to Mandelbrot's results: forswear any and all use of unidentified coastline interpretations. One way to avoid trouble is simply to consider coastline similarities as not evidence. This movement has gained wide adherence and has provided a secure area in which to do research on other aspects of historical cartography. Students in historical cartography are indoctrinated with this attitude, and it is often promoted with zeal. In fact, there is a widespread attitude that any work that interprets unidentified coastlines must *necessarily* be wrong.

The doctrine of forswearing coastline interpretation does not yield an ideal logical discriminator in the class of Occam's razor. The highly respected historical geographer J. Brian Harley declined to join in a blanket rejection of topographic evidence.[61] Mandelbrot himself stated, in a different but still relevant context, "The eye has enormous powers of integration and discrimination. True, the eye sometimes sees spurious relationships which statistical analysis later negates, but this problem arises mostly in areas of science where samples are very small."[62] We will be looking at a large number of samples. It is quite possible that in looking away from some coastline similarity one may be throwing out the historical-insight baby with the doctrinaire bathwater. An example can be found in Alfred Wegener's theory of continental drift. It was inspired by the matching shape of the coastlines of Africa and South America as well as other coastline correspondences.[63] Might there be some more-discriminating methodological principle than looking away that could at the same time provide a barrier against gross errors while yet allowing *strong* coastline similarities as evidence?

The real trouble with Hapgood's and Gallez's work is not so much that they used map shapes as evidence but that they drove the evidence to multiple implausible and unwarranted conclusions. Scientific research, as distinguished from logical deduction, actually has few hard and fast rules, but it does have many rules of thumb that guide researchers toward logically tenable results. The doctrine of disavowing shape correspondences is an attempt at such a rule of thumb. Besides disregarding that rule, Hapgood and Gallez violated one rule that astrophysicist Richard McCray calls the "tooth fairy rule," which states that a credible theory can invoke a mysterious, unknown agent ("tooth-fairy") once, but only once.[64] The above-mentioned authors' works contained a multitude of "tooth fairies." In Wegener's case there was a single mysterious, unknown agent—the driving force behind continental drift. That theory was subjected to huge controversy in its time, including claims of accidental shape correspondence. But it was grudgingly accepted as correct when investigators discovered the corroborating evidence of sea-floor spreading a half-century later and developed the theory under the name *plate tectonics,* and within that discipline, shape correspondence analyses are alive and well today.[*]

I will adopt a shape correspondence rule here that requires meeting five criteria:

1. Study only large, extended coastlines (not isolated minor wiggles).
2. Candidate coastal shapes must have multiple discernible features.
3. The features must be of appropriate proportions and relationships.
4. Only limited scale adjustments may be involved.
5. Unless very specifically justified, rotations will not be allowed.

Undue attention to minor features in single sources has been the cause of a large part of the mischief in the past; I stress overall shape and multiple cor-roboration. Only self-evident similarities will be used, avoiding as much as possible any controversy over whether one shape really looks like another. Some attempt will nevertheless be made to use objective correspondence methods that incorporate computerized matching. Furthermore, the passing of these criteria must lead to no implausible or mysterious agents or impli-cations whatsoever. The only unknown agent will be the postulate of trans-fer of Eskimo geographical information to Europe via the Norsemen. That plausible hypothesis is hardly a "tooth fairy" of the same magnitude that con-tinental drift was in Wegener's time. I endeavor to supply a plausible basis for every other implication of a given interpretation.

Wegener's arguments were not based purely on coastline similarity. He also found nongraphical support in the distribution of related flora and fauna on each side of the split as well as interior mountain range correspondences. We will look at nongraphical evidence here too, particularly in supporting tex-tual passages.

My approach will attempt to formalize the testing of the hypothesis by modeling the structure of a hypothetico-deductive system. In this approach,

---

[*]The detailed mechanism of seafloor spreading remains a "tooth fairy" even now, but there is absolutely no question that Wegener was right.

rather than the classical deductive paradigm used in historical research, the paradigm is that used in contemporary scientific research:

Hypothesis implies observation;
observation is true;
thus hypothesis is probably true.

The degree of confidence in this probability is increased by increasing the number of confirmatory observations. This paradigm has sparked its share of philosophical controversy, but in the real world it has proven highly useful and applicable.[65] In a softened form, with strict mathematical implication replaced by expectation, it has been widely used in the softer sciences, like archaeology. In our situation, the observations will be the trivial case of instantiation, so I provide the largest possible number of independent instances expectable from and consistent with the hypothesis that Eskimo geographical information about America passed through the Greenland Norsemen and into Europe. At the same time, I hope not to encounter any contradictory observations and to be prepared to modify my hypothesis to accommodate any that may arise. I at times apply this paradigm in local arguments for auxiliary hypotheses as well as overall. In the end, this paradigm is merely a formalization of the familiar colloquial detectives' observation that, "when you're on the right track, things tend to fall into place."

Nevertheless, the reader must realize that, even with a formally grounded approach, the ideas presented will also include speculations and conjectures. That continues to be the case with studies on continental drift,[66] and has been the case for most of the studies of the Norse encounter with America. Here, speculations will be clearly identifiable as such, will stay close to evidence, and will be surrounded by conservative examination. And no complex structures will be built on speculative bases.

Certainly any conservative overview of the maps analyzed here must allow that any of the correspondences of shape noted with North American land *could* be accidental. We shall see dozens of them, and a difficult problem is whether such correspondences are *all* likely to be accidental simultaneously. When the correspondences all lie in the same area of the world, all are subject to the same paradigms, and the selection of source maps is dictated by the entire body of available evidence rather than being arbitrary, the probability of accidental coincidence decreases substantially. We will see many of these correspondences corroborating one another, to build a circumstantial case for correctness of the approach. Readers who can withhold a reflexive disapproval of all correspondence arguments a priori will see that the results here are plausible, do corroborate one another, and are corroborated by topologically separate (and separately weighted) interior features and other nongraphical evidence. Readers may then decide for themselves if accidental coincidence seems, while possible, likely or not.

I

Outstanding
Misunderstandings

CLAUDIUS CLAVUS

IN 1406, SOUTHERN EUROPEAN SCHOLARS made the first Latin translation of the ancient Greek geographer Ptolemy's work. In the second century A.D., Ptolemy had perfected an advanced scheme of cartography from his predecessor Marinus of Tyre, one that allowed flat maps to account for the curvature of the earth. Ptolemy also avoided the decay of accuracy over generations of map copying by attaching tables of numerical coordinates for latitude and longitude for each mapped location, an early example of the fidelity of digital reproduction. His cartographic work, which was in addition to his theoretical treatises, may have included an atlas containing a world map and many local maps, all with coordinate tables.[1] The rekindling of interest in Ptolemy in the Renaissance ultimately would sweep away many of the problems associated with medieval cartography. More and more Renaissance scholars began to read his work, and among the earlier to do so was the French cardinal, Gulielmus Filiastrus. Filiastrus had someone make him a Latin translation of the map captions in a Greek atlas presumed to have been descended from Ptolemy's. The Latinized maps were appended to a 1407 translation of Ptolemy's theoretical work and handed over to Filiastrus in 1427. Filiastrus dedicated the work to the pope and took the manuscript home with him to Nancy, France. Following his death the next year, it remained there apparently without influence on other cartographers until its rediscovery in 1835.

This Filiastrus Ptolemy of 1427 is the earliest known copy to have any modifications or additions to Ptolemy's original maps and texts. Specifically, besides the Latinization, it contains a map (Figure 32) and a textual description of the Scandinavian regions. Scandinavia had been vaguely known to Ptolemy and even his predecessors Pliny and Mela, but not with enough detail that he could include it in his local chorographic maps. Early in the co-ordinate table's listing the source of the map is made clear.[2] Namely, while describing in great detail the small islands between Denmark and Scandinavia, the list's text includes the description: "in the South, Salinghi, on which is located Salinga, the native home of Claudius Clavus Suarto, the son of Niels Petrus Tucho and Margaret Ingrid Cicilia Osee Strango Vininngh, chorographer of these present parts, at 40° longitude, 57° 30′ latitude."

Claudius Clavus Niger Cymbricus is the Latinized name of a Dane who may locally have been known as Klaus Schwartz. He was an educated traveler who seems to have appeared in Rome around 1424. Clavus apparently was a visitor at the Curia of Pope Martin V in 1425. It is perhaps there that his work became known to Cardinal Filiastrus. In his time, Clavus came to be considered the leading authority on the geography of the North—indeed, the only authority. While modern scholars continue such plaudits, it will become apparent that Clavus was not actually trained as a geographer. In fact, I show that Clavus was *not* actually an authority on the North, the Scandinavian geographical vacuum remaining until more than a century later, to be dispelled by one Jacob Ziegler and his contemporaries. Clavus left no known record of the sources for his geographical ideas. Practically all that is known of Clavus has been reconstructed from secondary references who cite Clavus as their authority.[3] From these references it seems that Clavus did not take his ideas of the North from a lone source. Rather, his ideas evolved as he himself discovered more sources. I use the word *discovered* because Clavus apparently belonged to that group of humanist scholars who, like Poggio, took great delight in searching libraries for forgotten books and obscure manuscripts. I suggest, as an auxiliary hypothesis, that it must have been some such obscure information that Clavus, interpreting and lending undue authority thereto, thrust into the vacuum of Scandinavian cartography. The information was not necessarily particularly old, just esoteric. Support for this hypothesis will be forthcoming.

Not only was Clavus's map the first chorographic addition to Ptolemy, but it was the first known map to claim to show Greenland graphically. (The toponym *Greenland* had appeared on earlier maps, but no outline.) Mind you, this earliest map to show Greenland's outline did not appear until a decade after Greenland's settlements' supposed extinction. So, what was Clavus's source of data? Was it strictly conjectural? It would be a far stretch to think the outlines on this map could have been derived from knowledge of either Greenland or Scandinavia itself (see Figure 32 and front map). We shall see that, while an element of conjecture was involved, there was probably an actual source of geographical data about some "somewhere" transmitted via the Greenlanders.

The first step of interpretation necessary is to note the unreal arc of land that the map shows connecting Greenland to the Eurasian continent at the base of Scandinavia. Other writers have attempted to explain this in various ways. To some this arc looks as though it could represent the edge of the polar ice pack. Nansen believed that this representation involved some misconstruction of the leftmost headland of the Medici Atlas of 1351 (Figure 25), perhaps one of Clavus's sources.[4] But there is also a possible reality-based way to explain Clavus's unreal arc (as well as the Medici's). In reality, northern Greenland does come into proximity with other land: the Arctic Archipelago north of Canada, including Ellesmere Island. And this archipelago does come into close proximity with the continental Canadian coast. In fact, these proximities, bridged by winter ice, are exactly what made possible the Thule migration into northern Greenland from Alaska. We will see below that Clavus had information about that migration. Any description of that migration would necessarily include some suggestion of vague connection between northernmost Greenland, where the crossover occurred, and continental land. But Greenland lay far to the west of Europe. Under the one-ocean paradigm, Clavus was evidently unwilling to consider it *so* far west as to spring from eastern Asia and was forced to promote the hypothetical arc connecting Greenland to northeastern Eurasia somewhere beyond Scandinavia.

Many writers have unquestioningly followed the modern Scandinavian scholar Axel Björnbo's assertion that this conjectured arc was a fundamental part of all Scandinavian geographical theory. It was widespread, but I attempt, following Andrew Fossum,[5] to show later that there were several other geographical conceptions existent in the North.

Clavus himself admits indirectly, through his coordinate table, that the arc connecting Greenland to Eurasia is essentially conjectural. Namely, he lists specific coordinate data for every other place on the map except this arc. After describing three documented promontories of Greenland at 71° North (by the left-hand scale), Clavus's list terminates. In his textual chorography, he gives the coordinates of those first three headlands of Greenland and then, after the third, says: "But from this headland an immense country is extended eastward as far as outermost Russia. The pagan Karelians occupy the north of it, and their territory is extended past the North Pole to the eastern Seres [China], and therefore the pole which to us is in the north, is to them in the south at 66°." Clavus's belief that the North Pole could be in the north for some and in the south for others shows that he was not a trained geographer. He uncritically accepted an old gaffe of Marco Polo's, apparently based on a confusion between the North Pole and the Arctic Circle, in fact at approximately 66° north latitude.[*]

Still more anomaly appears in Clavus's mention of the "pagan Karelians" as occupying northern Greenland. The Karelians were actually a Finno-Ugrian people who inhabited the regions around the White Sea. In times past they had indeed been quite resistant to attempts at conversion to Christianity, but somewhat before Clavus's time they had been completely Christianized. This "pagan Karelian" anomaly has already been explained by Joseph Fischer and other authors as an example of the provenance paradigm, without using that term, putting the Karelians in the Eskimos' place.[6] This is simply a variant of the Eskimos' identification as Hyperboreans. We will see further development of the idea that Clavus was aware of the Thule Eskimos and their migration, even though misidentifying them as Karelians.

An alternative interpretation regarding this Karelian anomaly has been developed by linguists over the past century. As noted before, the Greenland Eskimos absorbed certain Norse loan words into their vocabulary. The Norse word for the Eskimos was *Skraelings.* Under the same pronunciation-shift paradigms for the other loan words, *Skraeling* becomes something close to *Kareli* in Eskimo,[7] and it may be the origin of the Greenland Eskimos' alternate name for themselves as *Kalali* instead of *Inuit.* In either case, *Karelians* as used by Clavus refers to Thule Eskimos.

The hard data in the rest of Clavus's coordinate table, as reflected in the map, will provide us with our first tangible evidence of the cartographic misunderstandings suggested. Let us focus our attention on the strange graphical representation that Clavus identifies as Scandinavia (refer to Figure 32). The first step of analysis for this or any other map we shall look at will be to deconstruct the map, using this word in a slightly different form of what was advocated by the historical geographer Brian Harley.[8] He argued that maps constitute a cultural text akin to verbal texts in that they can be analyzed for far more information than just their surface appearance. However, instead of looking for political or social ideologies, we will be extracting alternative graphical dualities that escaped the creator's conscious mind. When a cartographer puts together a new map, he has many things in his mind. The

---

[*]Apologists for Clavus tried to explain this away with convoluted arguments. Apparently they were unaware that the idea came straight from Marco Polo (Chapter 2 of *The Book of Marco Polo,* "Road to Cathay").

foremost is usually some preexisting picture given by other cartographers. Almost all new maps incorporate parts of previous maps, to which alterations or additions are made. Attempting to unveil the rationale of those alterations and additions can give a glimpse into the cartographer's mind and less obvious sources. In reconstructing the map while analyzing the hypothetical influences on the cartographer, we will always start by focusing on the graphical essence that was new in the map at its time and has no known source in reality.

To that end, we should *not* be focusing, in Clavus's map, on the known lands of Denmark, England, and Ireland, which were all copied from Ptolemy. We should also *not* be focusing on the conjectural northern Greenland coastline arcing to Asia, for the very reason that it is conjectural and undocumented in the coordinate table. Let us also, for the time being, not focus on the toponymic cities and landmark identifications, which are not graphical and which probably would have come from an originally separate data resource. What remains that is cartographically novel, including the three quantitatively specified data points or promontories of Greenland from the coordinate table, is shown in the top of Figure 3. This is the new cartographic or graphical text, whether hard copy or descriptive or imaginary, that Clavus, from whatever sources and for whatever reasons, fitted into the existing chorography of northwesternmost Europe.

The bottom of the figure shows the accurate graphical chorography of northwesternmost America, a detailed map of the Bering Strait and Seward Peninsula. The resemblance is striking, sufficiently so to suggest an outrageous hypothesis. It appears that this could be a candidate for a misunderstanding that placed Alaska's chorography in Scandinavia. Clavus's "Gronlandia Provincia," the remaining three data points when the conjectural upper coastline is removed, could be interpreted as the tip of the Chukchi Peninsula of Siberia; his "Islandia" occupies the same position as the Diomede Islands; and his would-be Scandinavia nearly perfectly represents the actual Seward Peninsula. These Eskimo lands bear the same geographical interrelationships as the Scandinavian ones. As we proceed, we will see evidence to make the above hypothesis seem less and less outrageous.

The area around Bering Strait has always been of great significance in Eskimo history. It and nearby areas are considered to be the birthplace of many Eskimo cultures, including the Thule.[*9] Is it plausible to think that Thule

----

*While the name archaeologists give to a culture comes from the site of its first modern unearthing, this has no relation to the historical site of its first development. That is determined from extensive further excavations and analysis. Early in the history of Thule excavations the site

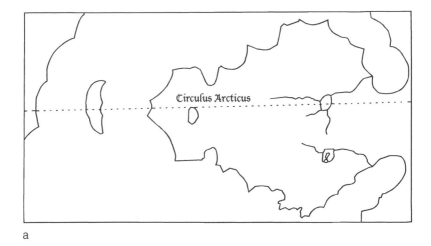

a

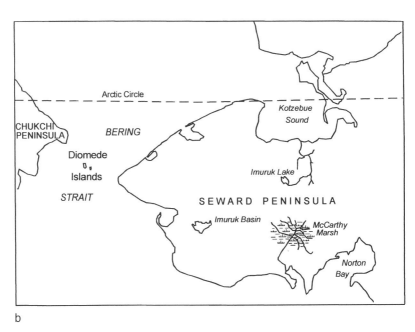

b

Figure 3. Portion of 1427 map purportedly of Scandinavia by Claudius Clavus (*a*) compared with Alaska's Seward Peninsula area (*b*)

cartographic information about this area could have made its way to the Norsemen in Greenland and thence to Clavus in Europe? One guide to making this interpretation might be sought in the historic Eskimo geographical

of origin seemed to be Point Barrow on the north coast of Alaska. But Thule sites just as early or earlier have been found along the Bering Sea south of Seward Peninsula. This highlights the intense communication and mobility of the predecessor culture. The Thule explosion grew out of the Birnirk Culture, which had settlements in Seward Peninsula.

interaction with early modern explorers. The historic Inuit Eskimos are direct cultural and genetic descendants of the Thules. Such early modern explorers found authoritative Eskimo guidance widely available at many points.[10] One of the early explorers in this Bering Strait area, Frederick Beechey, encountered a party of local Eskimos, who of course shared little if any language with him and must have communicated primarily by gestures. They were in the act of drawing in the sand exactly such a chart of Seward Peninsula and Bering Strait that included the Diomede Islands and the tip of the Chukchi Peninsula:

> On the first visit to this party, they constructed a chart of the coast upon the sand, of which I took very little notice at the time. Today, however, they renewed their labor, and performed their work upon the sandy beach in a very ingenious and intelligible manner. The coast line was first marked out with a stick, and the distances regulated by the day's journeys. The hills and ranges of mountains were next shown by elevations of sand or stone, and the islands represented by heaps of pebbles, their proportions being duly attended to. As the work proceeded, some of the bystanders occasionally suggested alterations, and I removed one of the Diomede Islands which was misplaced: this was at first objected to by the hydrographer; but one of the party recollecting that the islands were seen *in one* from Cape Prince of Wales confirmed its new position, and made the mistake quite evident to the others, who seemed much surprised that we should have any knowledge of such things. When the mountains and islands were erected, the villages and fishing stations were marked by a number of sticks placed upright, in imitation of those which are put up on the coast wherever these people fix their abode. In time we had a complete topographical plan of the coast from Point Darby to Cape Krusenstern. . . .
>
> . . . For East Cape they had no name, and they had no knowledge of any other part of the Asiatic coast. Neither Schismareff Bay nor the inlet in the Bay of Good Hope [both in Alaska] was delineated by them, though they were not ignorant of the former when it was pointed out to them.[11]

So Beechey's Eskimos' geographical knowledge covered exactly the area shown in the bottom of Figure 3, albeit with some imperfections. Rundstrom has noted that in Inuit cartography it is the activity rather than the final object that is important. This particular group was having an animated discussion as they proceeded, correcting one another's contributions to the drawing. They carried out their activity on more than one day. Clearly this activity constitutes a group learning experience, if not exactly a classroom.

Perhaps this is why they were surprised that untutored foreigners would have any knowledge of such things. Australian Aboriginal mapmaking had a similar emphasis on the importance of the knowledge reinforcement rather than a resultant artifact. While they did ceremonially make hard copy maps, there was no emphasis on preserving them, and sometimes they were intentionally destroyed afterward.* Before the invention of the printing press, people employed mental practices that yielded feats of memory that seem phenomenal to us today.[12]

The next year, Beechey returned to Bering Strait and sent a crewman exploring in a slightly different area: "He was visited by several of the natives while there, one of whom drew him a chart, which corresponded with that constructed upon the sand in Kotzebue Sound the preceding year.[13] So this picture of Bering Strait seems to have been a not so unique accoutrement in the area. It is plausible that a medieval Thule migrating eastward from there could have carried such a picture in his head. Would it be plausible for him to migrate all the way to Greenland?

The most widely traveled Eskimo recorded in historic times is Qillaq, who led his family and others ultimately from southern Baffin Island to northern Greenland.[14] Their actual transit through Devon and Ellesmere Islands took less than three years, even though they were traveling through uninhabited territory, carrying all their possessions. One of the most remarked upon aspects of the medieval Thule migration is the rapidity with which it populated the American Arctic,[15] so it is not physically impossible that a late medieval Eskimo with adult knowledge of the Bering Strait could have migrated east, with the aid of alliances in earlier Thule settlements all along the way,[16] and carried his geographical knowledge in his lifetime to northern Greenland. There, he would have encountered ample risk of being kidnapped to Europe and interrogated about his origins.

One area of likely contact between the medieval Norsemen and medieval Eskimos is in what archaeologists call the "Gateway to Greenland," between Kane Basin and Hall Basin on the northwestern shore (see Figure 5). This is where the migration crossed over the ice from the Arctic Archipelago to Greenland itself. A particularly likely contact site, of the many possible, is tiny Ruin Island off the shore of Inglefield Land. Here archaeologists found Norse cultural trinkets throughout Eskimo remains in the ruins. Recent work shows that the Eskimos who lived here constituted a later migration of

---

*Those that have incidentally been preserved to modern times have taken on a singular importance to native land claims against the modern government.

Thules, whose culture did not derive from the north coast of Alaska as did that of the first Thules. Instead, it was very much like the "Western Thule" culture that occupied exactly the area shown at the bottom of Figure 3.[17] These cultural affinities were so strong that they suggest a very rapid, direct migration somewhat akin to that of Qillaq and his group.[18] This kind of situation could very plausibly have led to cartographic information about Seward Peninsula reaching the Norsemen.

Lest there be some possibility that the coastline similarities in Figure 3 could be just cartographic coincidences, let us examine more detail. Clavus shows two disparate scales of latitude on his map, a left-hand scale and a right-hand scale. Filiastrus or his draftsman has identified the left-hand scale as due to Ptolemy and the right as due to Clavus. But Ptolemy had no detailed knowledge whatsoever of the Scandinavian region, and the attribution of such a scale to him is without basis. The right-hand scale is correct for the latitudes of England, Ireland, Denmark, and the place names in Scandinavia. Surprisingly, it is not used in the text.[*] The left-hand scale is the only one consistent with the graphic and the text's latitudes, even though it is quite inaccurate for all the named locations. However, it is nearly perfect for the Bering Strait area. In fact, the Arctic Circle, 66° by the left-hand scale, actually does come within half a degree of being the true axis of symmetry of Seward Peninsula.

This scale provides a basis for a still stronger corroboration of such an interpretation of the map. Toward the right end of Clavus's would-be Scandinavia, barely above and on the 66° line (by the left-hand scale), he shows an inland lake. This corresponds closely with Imuruk Lake in Alaska, just above the axis of symmetry of Seward Peninsula. Closer to the tip of the peninsula and just below the axis, he shows a lake which corresponds to Imuruk Basin. Finally, exactly south of the first lake and quite below the axis of symmetry he seems to show a lake with interior islands part way up a river. This may have been a good representation of the Fish River system, which forms a focus today with the thirty-mile-diameter McCarthy Marsh at its center. It may well have been wetter then, with five fewer centuries of isostatic glacial rebound. There are today a number of low hills in the marsh that would have been islands. At any rate, lake or marsh, it would have been just as impassable then unless frozen and would deserve an appearance on the map. It is

---

[*]More than a century later cartographers were to use double latitude scales in an interim attempt to deal with compass variation. The amount of offset here seems inappropriate to serve such a purpose, and there was no practical reason for such an idea to be in existence so early, before sailors on the Atlantic encountered serious variation.

difficult to find any interior features in the real Scandinavia that as a group correspond with the map so well as these. Such interior features constitute evidence that is topologically separate from the coastline similarity and should weigh independently and additively.

It would seem plausible that the left-hand scale could actually spring from some indication accompanying original data from the Bering Strait area. Obviously that would not have included degrees of latitude, which were introduced in Europe only with the widespread acceptance of Ptolemy, and before which the number of hours of daylight on the longest day was the standard expression of the concept of latitude. The only measurement necessary would have been the location of the Arctic Circle daylight transition. Because the Arctic Circle transition is detectable without any instruments or theoretical concepts, a European cartographer interviewing a Hyperborean (or "Karelian") could have found out if he came from above or below the Circle just by asking "Did they lose the sun in winter where you were?" The conversion in Europe of any such indication into degrees of latitude, its scaled extension northward and southward, and its attribution to Ptolemy would be natural for the era. That attribution could have been reinforced by a well-known error of Ptolemy's own in the latitude of Britain.

Clavus's own scale, on the right, would have been necessitated by his inability to reconcile these hypothetically Alaskan latitudes with known latitudes of cities in the real Scandinavia and other parts of northern Europe. The size of Filiastrus's and Clavus's degree, however, has enlarged the relative size of Seward Peninsula by a factor of around ten. Certainly the Eskimos had no concept in their maps of scale compared to Europe. That Clavus was somehow aware of this disparity and found it unpalatable is suggested by changes he made in a later map we will examine below. The longitude scale (properly reflecting Ptolemy's conic projection with smaller longitude increments here than of latitude) also corroborates the exaggeration. It gives a distance between western Ireland and Copenhagen that is more than 40 percent greater than the true value. This is all consistent with an attempt to force the Ptolemaic grid over a preexisting graphic.

Waldo Tobler has developed a suite of computer programs useful in comparing supposedly similar two-dimensional figures, whether they be missing person sketches or old maps.[19] One of its most fundamental results is a correlation coefficient measuring the best degree of fit possible between two images under rigid rotation, translation, and scale change. Applying this program to Figure 3 yields a correlation coefficient of 83 percent.* Applying the same procedure to the comparison in Figure 1 results in a correlation coefficient of 75 percent. What this means is that, objectively, the Clavus map "of Scandinavia" is an even better map of the Bering Strait than the Inuit map "of Southampton Island" is of Southampton Island.

It is still theoretically possible, of course, that both the coastline and interior similarities arose by chance, even if a remote chance. One way to reduce the likelihood of that is to find circumstances regarding the map, other than the graphic itself, that tie it to the Bering Strait or the Eskimos. Such circumstances arise in the verbal text of the coordinate table. The suggested Bering Strait prototype has a feature that could have fed Clavus's mental struggle about Greenland's northern extremity. Hypothesize that the three quantified data points of the map's would-be Greenland do originate in the Chukchi Peninsula. Then Clavus's statement would be true, that they are connected in a sweeping arc to outermost Russia. namely via the Asian Arctic coast. Remember that Beechey's Eskimos did have a vague knowledge that what we call Asia was there. Before the rise of the Soviet Union, Eskimos were known to transit regularly between Alaska and Asia (via ice and the Diomede Islands). The Bering Strait has been an active trade link for two thousand years, and there has always been a great deal of communication among the settlements.[20] This is obviously not to suggest that the medieval Eskimos knew about the country Russia or the continent Asia, but a verbal or drawn description of the realities could have led to jumping to such a conclusion in Europe.

A SECOND MAP OF CLAVUS'S IS PRESUMED to have been made shortly after his first, ca. 1431. We will see that information associated with it further reduces the likelihood of accident in the first map's resemblance to Bering Strait. No copies of the second map remain in existence for us to check, but by a very fortunate combination of circumstances we are able to know al-

---

*Certain underlying assumptions are necessary to make this calculation. Not the least of these assumptions is deciding which points or features in the two images seem to associate with one another. As a first step, note that the Clavus map actually contains many stylized contours. The decision was made to use only the points specified in the numerical coordinate text, which are the cusps in the coastal outline. Furthermore, since the computer program prefers somewhat evenly spaced points, some of the densely spaced points in the southeast were omitted. Another decision was to ignore Clavus's suspect longitudes entirely, basing the comparison points on an arbitrary rectangular grid. After all, it is strictly the graphic aspect of the map that we want to compare. Detailed documentation of the calculation may be had from the author.

most exactly what it must have looked like. Axel Björnbo discovered a copy of a Ptolemaic style of coordinate table placed by some Renaissance scholar between the pages of a book still existing in Vienna. Björnbo immediately recognized that it must have been a copy of the coordinate table behind Clavus's second map.[21] Its first words are: "I, Claudius Clavus Niger the Dane, son of Nicholas Petrus Tuco and Margaret Christiana Strango Vinninch, have taken pains to preserve faithfully for posterity by a drawing and textual record, the following acknowledged regions, known to me with mathematical accuracy by actual observation, which were unknown to Ptolemy, Hipparchus and Marinus." Danish writers early in the twentieth century, understandably eager to convince themselves that the first surveyor and cartographer of the North was a Dane, have interpreted this testimonial to mean that Clavus physically went to and surveyed all these places. But this seems unlikely to be correct, especially considering Clavus's demonstrated limitation as a geographer. Recall that the Norse had no cartographic tradition. The declaration may just as well suggest that Clavus the collector knew of someone else's map or information which arose from observation and he preserved it for posterity. This suggestion will be supported below.

Björnbo set about using the coordinates in the textual record to regenerate the lost map. The map he came up with proved to be essentially the progenitor of most of the known maps of Scandinavia drawn in Ptolmaic atlases for the following century![22] The early map of Donnus Nicholaus Germanus, in the Zamoiski Codex (Figure 43), must be an exact copy of the coastlines of Clavus's second map. For now, we shall speak of it as though it were Clavus's map.

This second map of Clavus's shows marked evidence of a discovery of new information. He might have been dissatisfied with the first map on at least two counts: first, the land bridge to "Greenland" around the North was confusing in the asserted prototype; and second, the exaggeration of scale mentioned above might have somehow made itself known.

His dissatisfaction with the first map would not have been with its degree of likeness to Scandinavia, which Clavus couldn't have known. The second map is no better in this respect than the first—indeed, it is worse. He presumably turned to a new source because it did not have the drawbacks of the first and it seemed to fit reasonably well into the known map of Europe. Can we find an area in Eskimo territory whose transformation under one of the divulgence-hiding paradigms could lead to this map?

A candidate for this map's prototype, again outrageous at first encounter,

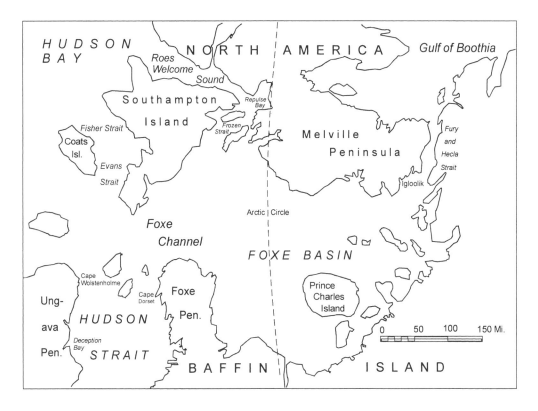

Figure 4. Sideways (to modern eyes) view of Foxe Basin and Southampton Island area

is shown in Figure 4. This is a *sideways* (in modern eyes) view of the area around Southampton Island in the entrance to Hudson Bay (see front map). I posit that Clavus received (directly or indirectly) a native map of this area from the hands (or mouth, allowing a descriptive source) of someone coming through Scandinavia and took it to be a map of the Scandinavian region—an instance of the provenance paradigm. The identification of the European counterparts under the suggested misunderstanding is straightforward. Foxe Basin becomes the Baltic Sea, Foxe Channel becomes the Skagerrak and Kattegat, Foxe Peninsula becomes the Danish Peninsula, Hudson Strait becomes the North Sea and the tip of the Ungava Peninsula becomes Great Britain. (I am not necessarily suggesting that there was some such prototype in its entirety but am demonstrating how one could have accommodated to and assimilated any part of it.) This accounts for all the features on Ptolemy's original map of the area. Clavus had no reason to change these features from their Ptolemaic representation, but he would have been justified in deciding that any such known likenesses were good enough for him to trust Southampton Island as a candidate for Scandinavia. Even without any computer analysis* one can see the general similarity between this Clavus "Scandinavia" and the top of Figure 1. (Remember that the left side there corresponds to the top of Clavus's.) An even greater similarity exists in the Inuit map of Southampton Island there.

One can now see that the (unreal) channel between Clavus's Scandinavia and the mainland would in this model have had its genesis in the real Roes Welcome Sound between Southampton Island and the Canadian mainland.† Would-be Iceland on the second Clavus map, as surviving through Donnus Nicholaus, is drawn stylistically as on the first but seems to be a passable representation of Coats Island, perhaps somewhat misplaced.

Consider that one of the posited reasons for Clavus's switching prototypes had to do with scale. He would have had to change the absolute scale of this map by a factor of less than two to make it fit comfortably with Europe. That is a great improvement over the alleged scale distortion in the first map and is consistent with other known early scale disparities that are widespread in the history of cartography, some of which we will later see.

However, one glaring doubt must remain in our minds. To fit the prototype to Europe we have had to make its west point north. This problem is only illusory, however. It probably would not have confronted Clavus at all. In fact, the Eskimos did not have any preferred or absolute orientation for their maps and did not have the European global concepts of north, east, west, and south. These concepts arose in most cultures from the simple orderly behavior of the Sun, missing in Arctic latitudes.[23] The Eskimos' cardinal directions were based on major coastline trends and prevailing winds, which could vary from one major locality to another.[24] In particular, the cardinal word *uangnaq,* defining our west, at mainland Igloolik points to our north in Southampton Island usage. Eskimos talking to one another had ways to deal with such shifts.[25] Communication between a Hyperborean (or "Karelian") and a European cartographer would have been less clear. A resulting rotation of orientation in the observed direction and amount is quite plausible.

---

*There will be dozens of map comparisons to examine herein, and developing a computer generated correlation coefficient for each would serve little purpose at substantial effort. Insufficiency of the statistical approach to coastline similarity has been demonstrated by the case of Charles Hapgood, who achieved high correlations that are spurious. The eye recognizes easily enough the fact of similarity in all of the cases we will examine, and the real question to address is whether they are accidental or meaningful. We will channel our efforts instead toward that question.
†Other maps soon evolved isthmuses bridging the channel. See discussion on page 148 ff.

The information putatively about Southampton Island in Clavus's second map and coordinate table has further, nongraphical, detail. In that table, in labeling the northernmost promontory, Ynesegh, of his would-be Scandinavia, Clavus calls it "Ynesegh promontory and market town—where the frozen sea called *Nordhenbodhn* itself, by means of a long water tract, enters the Sea of Gotland [Baltic] towards the Northeast—whose location is 41° Longitude 66° Latitude, the Arctic Circle," thereby unequivocally reinstating a long-abandoned theory of an insular Scandinavia. One reason he may have had for doing this was that Southampton Island is an island, separated from the mainland by just such a water tract as described—Roes Welcome. In fact, the northernmost promontory of Southampton Island is almost exactly at 66° north latitude. The Norsemen apparently did trade with the Eskimos,[26] and the northernmost tip of Southampton Island, right below the Arctic Circle, would have been a mutually easily definable place to meet for such activity. The now extinct tribe of Eskimos from Southampton Island, the Sadlermiut, always seemed to anthropologists to be strangely different culturally from other known Eskimos. Perhaps, one might speculate, Norse contact influenced their culture.

The curious frozen sea that Clavus calls "Nordhenbodhn" (properly *Nordnbotn,* the Bayhead or "bottom" of the North) has appeared in other Norse geographical references. It has always been a problem to fit into the geographical scheme of Scandinavia without resorting to fantasy. One can now see a reason why, for the text here evidently refers at least in this context to the fact that Hudson Bay connects with Foxe Basin via Roes Welcome Sound. We will see evidence later for an awareness of Hudson Bay. The map itself in the Zamoiski Codex does not use the label *Nordhenbodhn,* but it does show a "Hyperborean Sea" to the left of the Southampton and Coats Islands analogues.

Somewhat later in his coordinate table Clavus defines the coordinates of a location which, relative to the identification of Ynesegh as the northernmost cape of Southampton Island, would correspond to the east coast of Melville Peninsula. Here the continental Arctic coast begins (see Figure 4). Of this coordinate he says:

Uttermost limit marked with a crucifix, so that Christians shall not venture to go beyond without the king's permission, even with a great company; located at 43° longitude 67° latitude. And from this place westwards over a very great extent of land dwell first the Wildlappmanni, who are thoroughly wild and hairy people, as they are depicted [evidently Clavus's original, unlike Donnus Nicholaus's copy, was a pictorial or illustrated map]; and they pay yearly tribute to the king. And after them farther to the west are the little Pygmies [Eskimos], a cubit high, whom I have seen after they were taken at sea in a little hide-boat [kayak], which is now hanging in the cathedral at Nidaros; there is likewise a long vessel of hides [umiak], which was also once taken with such Pygmies in it.

Norway has eighteen islands, which in winter are always connected with the mainland, and are seldom separated from it, unless the summer is very warm. The peninsula of Greenland island stretches down from land on the north which is inaccessible or unknown on account of ice. Nevertheless, as I have seen [this doubtful claim is examined below], the pagan Karelians daily come to Greenland in a great flock, and that without doubt from the other side of the North Pole. Therefore the Ocean does not wash the limit of the continent under the Pole itself as all ancient authors have asserted.

Such a barrage of descriptive information (it even continues beyond here) is unique in geographical coordinate tables. It should raise a flag and prompt us to look for some explanation. One good reason for its occurrence would be the uniqueness of the information. If Clavus had it on good authority, it would behoove him to mention it in the table even if it had no justification for being on the map.[27] One supportable hypothesis is that his authority was the now-lost book *Inventio Fortunatae.* I show in the next chapter that this possibly comprised a travelogue of lands west of Greenland, including the central Canadian Arctic. If Clavus did plagiarize the *Inventio,* then his doubtful claim to have been in Greenland and seen the incoming "pagan Karelians" is probably a quotation of that book's anonymous author. Indeed, that book is known to have contained information (latitude measurements) "known with mathematical accuracy by actual observation," the phrase in Clavus's opening statement of the Vienna text.

Here Clavus ties the "pagan Karelians" directly to Greenland rather than to any hypothetical arc in the north. In late phases of the migration, the Eskimos did in fact occupy the northern part of Greenland and were resistant to known attempts at conversion. And at the time Clavus was writing, they were entering Greenland in great numbers. He tells us that without doubt they came from the other side of the North Pole, which was true of the Thule Eskimos.

Recall that Clavus's Iceland corresponds in this model to Coats Island off Southampton Island. In the coordinate table he states, "Tyle [Iceland, which was believed to be the 'Ultima Thule' of Pytheas[28]] is a part of Norway and is not reckoned as an island, although it is separated from the land by a channel or strait, for the ice connects it with the land for eight or nine months, and therefore it is reckoned as mainland." This statement corroborates the Coats Island hypothesis as follows: the real Iceland is certainly not separated from Norway by a strait that freezes solid for months (see front map ice line), but the statement is quite true of the relation between Coats Island and Southampton Island.

A second improvement posited for Clavus in changing his first map, besides the scale exaggeration, concerns the hypothetical land bridge connecting Greenland to the continent. Clearly the map of this new area contains nothing to give rise to the actual graphic depicting Greenland on the second map. Nevertheless the geography around the suggested prototype could have reinforced Clavus's land bridge idea. Namely, the mainland behind Resolution Island does curve out to the left and eventually, following the shore of Hudson Bay and Strait, returns in a sweeping arc to the Ungava Peninsula at the opposite side of Resolution Island (see front map).

The actual graphic that Clavus used for Greenland, as we shall see, had appeared and would reappear in the Scandinavian corner of many other cartographers' works. It appeared with varying identifications ranging from Greenland to Scandinavia itself. There does exist a feasible prototype for this graphic in Eskimo sources from the Alaska area, wherein it looks somewhat like the main Aleutian or Alaskan Peninsula, which projects southwestward from the mainland of Alaska. Examples we will see below have an even more striking resemblance. That real peninsula indeed fulfills Clavus's earlier description of Greenland's sweeping arc connecting to the continent. He just named the wrong continent. The peninsula does sweep to the east to make the connection, even though the real Greenland was known to lie west of Scandinavia.

Meanwhile, we note for future reference that Clavus has now somewhat retreated from his original claim that the Greenland known to him was connected by such an arc to Russia. In the text, he refers explicitly to Greenland as an island and is unsure about its northern origin amidst the ice and other lands, yet he continues to rationalize about how the "Karelians" must have gotten there. This is consistent with the Thule migration's island hopping in the Arctic Archipelago across ice bridges to northern Greenland.

There is one further line of evidence confirming that Clavus's maps of the North were based on received graphics rather than his own survey. Note how sparsely his first map is supplied with toponyms (Figure 32), showing only the well-known towns in Scandinavia. The second map and its descendants (Figure 43) appear to compensate for this, being filled with toponyms.

Björnbo has shown that this labeling is a complete fraud, the new names being mostly derived artificially from a medieval Danish folk song, the runic alphabet, and numerals, and from Danish dialectal children's doggerel.[29] The only plausible motive I can see for something like this is to lend authority to his received graphic with north-sounding names that his southern readers were unlikely to see through. This device was reasonably successful for almost five centuries.

Thus ends the known geographical legacy of Claudius Clavus. There is some tenuous evidence, later to be seen, that he produced a third map, now lost, also based on North American land. We may be thankful that he faithfully recorded rather than dismissed as fantasy that which he did not understand, for in doing so he may have given us the first plausible glimpse into Norse transmission of geographical information from west of Greenland.

CHAPTER 3

THE *INVENTIO FORTUNATAE*

AND MARTIN BEHAIM

AS WE HAVE SURMISED, maps as early as Clavus's of 1431 might include data from the lost book *Inventio Fortunatae,* of about 1363. The 1507–8 map of Johann Ruysch (Figure 54) explicitly claims such data. It shows a ring of eighteen stylized islands surrounding the North Pole, and a nearby marginal inscription says that the data come from the *Inventio Fortunatae.* Gerard Mercator also showed a version of these eighteen islands, in his map of 1569 (Figure 68). The number eighteen resonates with Clavus's statement in the Vienna Text that Norway has eighteen ice-bound islands, and the correspondence strengthens the surmise that Clavus knew of the book. The 1427 Filiastrus version of Clavus's first map shows fourteen stylized islands distributed among each of the stylized bays of northern and western Scandinavia, two more, somewhat larger, symmetrically at the base, where Scandinavia joins the continent, and two symmetrically above and below the Arctic Circle between Scandinavia and Iceland, totaling eighteen.* This suggests that the draftsman may have been trying to accommodate to similar descriptive information based on the *Inventio Fortunatae.*†

The only knowledge today of the contents of the lost Inventio comes extremely indirectly through another lost book, the *Itinerary of all Asia, Africa and the North.* Written by the Dutchman Jacob Cnoyen of The Bosch ('sHer-

---

*Some line-reproductions of this map (e.g., Nordenskiold's and my own in *Viking America*) do not include these last two islands for a technological reason. The original map contains a color wash over its sea area, which for line-reproductions has been intentionally removed by photographing in high contrast. In the process, the somewhat weak outline of these islands has been inadvertently lost there.

†Some of these islands are identified with known islands in the Nancy text, but we have learned that such identification may not be authoritative.

togenbosch), it contained extracts from another lost work, *Gestae Arthuri,* intermixed with a summary of part of the *Inventio* as told by another intermediary.[1] This intermediary was an unnamed wandering priest who had acquired an astrolabe[*] directly from the unnamed author of the *Inventio* during face-to-face conversations. The famous Renaissance cartographer Gerard Mercator quoted verbatim parts of Cnoyen's now also lost work and summarized others in a letter to Dr. John Dee, and Dee made a transcript that survived after Mercator's original became lost. Dee's transcript itself is now slightly marred by fire damage. Mercator wrote his letter in response to Dee's questions about source material for the curious "Septentrional[†] Islands" on his map of 1569. This map, of which Figures 68a and b are details, is the one in which Mercator introduced his projection scheme.

Dee's text, to be examined in more detail later, includes a fantastic-sounding description of eighteen to twenty islands, the "Septentrional Isles," adjacent to the North Pole. They were separated from one another by nineteen "Indrawing Channels," which would suck any approaching ships into an exitless central sea or dash them against rocks.[2]

E. G. R. Taylor has pointed out that many of these features satisfy known medieval and earlier symbolic fantasies about the Far North.[3] Delno West showed further evidence that preexisting fantasies influenced this description.[4] All of these treatments start from a fundamental presumption that the *Inventio*'s description must be a complete fantasy. None of them attempts to prove this presumption; they merely look for possible roots of such assumed fantasy. The claims are not without merit, but nor are they overwhelmingly persuasive. Under the circumstances, it might be a mistake to assume that the *Inventio* contains no geographical reality at all. Mercator's map itself contains the following caption over the polar area:

> In the matter of the representation, we have taken it from the *Travels* of Jacob Cnoyen of the Buske, who quotes certain historical facts of Arthur the Briton but who gathered the most and the best information from a priest who served the king of Norway in the year of Grace 1364. . . . He related that in 1360 an English minor friar of Oxford who was a mathematician, reached these isles and . . .

---

[*]Astrolabes modeled the daily rotation of the star map throughout the seasons and included a graduated scale for measuring the elevation angle of a star above the horizon, yielding the observer's latitude.

[†]After the seven stars of either the Big Dipper or the Little Dipper, including Polaris, hence North Polar.

described all and measured the whole by means of an astrolabe somewhat in the form hereunder which we have reproduced from Jacob Cnoyen.

Many rounds of interpretation have intervened between the Minorite's description of "these isles" and what we see today, allowing much room for fantasy and stylization to enter. It is still possible that there actually were preserved some kind of measurements, "known with mathematical accuracy by actual observation" in Clavus's words. Furthermore, it would be an unjustified leap to conclude that Cnoyen actually read the *Inventio*. He did read the *Gestae Arthuri*. Even the priest with the astrolabe who spoke face to face with the *Inventio* author does not claim to have read it; he merely claims to relay the author's words. It would also be a leap to conclude that Cnoyen got any information directly from the priest. We have no information on when Cnoyen lived;[*] he only claims to rely what the priest reported to the Norwegian king, which could have reached Cnoyen through other sources. One circumstance weakly suggests he could have met the priest: the Belgian Cnoyen (Mercator recorded that he wrote in a "Belgic" tongue) reported the personal information that the priest was a fifth-generation Bruxellois. The map by Ruysch, which explicitly mentions *Inventio Fortunatae* in an inscription (Figure 54), is even more stylized than Mercator's. In fact, Ruysch does not actually claim *Inventio Fortunatae* as his source either. His use of the phrase "One may read in *Inventio Fortunatae* that . . ." does not necessarily mean that he had read the book. He may have been just quoting Cnoyen.

Let us weigh more closely the stylization appearing in these maps. Symmetry is always attractive to humans, and in the Renaissance there was a strong Neoplatonic movement involving symmetry. Scholars often turned to the principles of symmetry when other information was lacking. Clearly there is cartographic evolution visually centered on the North Pole from Ruysch to Mercator (and appearing in other cartographers during the interim) that could have been based on evolving interpretations of Cnoyen. Clavus's stylization of the eighteen islands is completely different, symmetrical about the Arctic Circle. Remember that Clavus adopted Marco Polo's gaffe, saying "the pole which to us is in the north, is to them ["Karelians"],

---

[*]The town archives of 'sHertogenbosch show that Cnoyen in various spellings was a common name in the fourteenth and fifteenth centuries, with at least two Jacobs on record in the fifteenth. Mercator received his early schooling in 'sHertogenbosch. At the end of his transcript, Mercator states that he copied it down "many years ago." This may (or may not) give a lead to Cnoyen's florient.

in the south at 66°." This could explain his use of symmetry with respect to the Arctic Circle. However, in reading Cnoyen's written description closely, it is difficult to find much emphasis on symmetry. Where the cartographers deploy their islands around the North Pole, Cnoyen only says they are "adjacent" to the pole.

In the Martin Behaim globe (Figure 48), the Septentrional Islands described verbally by Cnoyen are given a far less stylistic rendition. This globe was produced at Nuremberg in 1492 just before Columbus set sail; it also happens to be the oldest known terrestrial globe remaining in existence. We will refer to it throughout this chapter. Note that the islands are adjacent to, rather than surrounding, the pole. Although Behaim cites no sources for the circumpolar area of his globe, it is assumed that the source was the *Inventio Fortunatae*.[5] However, we have no information on whether the probable source was the original *Inventio* or Cnoyen's summary. Concerning his sources for the North, Behaim says only that "the far-off places toward midnight or Tramontana, beyond Ptolemy's description, such as Iceland, Norway and Russia, are likewise now known to us, and are visited annually by ships wherefore let none doubt the simple arrangement of the world, and that every part may be reached in ships, as is here to be seen."[6] Note that Greenland appears to be depicted twice on the Behaim globe.[7] The peninsular version of Greenland, which Behaim labels "Groenland," just north of Scandinavia is copied from so-called B-type maps descended from Claudius Clavus (to be studied later). I refer here instead to the uncaptioned island showing a picture of an archer and a bear, which I conjecture is a disproportionately fattened Greenland. In later chapters I will justify this conjecture.

Just west of this Greenland, nestled in the crook of its west coast, lies an island. This is probably the same island Mercator calls Grocland. Cnoyen's polar description includes a mysterious island he called Grocland, west of Greenland.[*] Mercator's rendition of Grocland, somewhat irregularly naturalistic in shape, looks quite realistic, in contrast with his stylized Septentrional Islands.

The naturalness of these renditions putatively of Greenland and the yet-to-be-identified Grocland on Behaim's globe leads to the question of whether the other islands adjacent to the pole could derive from reality. They too have an irregular naturalism. The verbal description of the Septentrional

---

[*]Dee's transcript of Mercator's letter contains a marginal gloss, presumably Dee's own, suggesting that Grocland is meant to be Greenland. This speculation is based only on the fact they both begin with *Gr,* and we will see that it is wrong.

Islands as preserved by Cnoyen does seem to contain many elements of fantasy, but it is also replete with details that could refer to actuality. I will provide an interpretation that allows for the reality of some of the surmisedly fantastic sources.

The only possible candidates for such an interpretation are the group of islands forming the Arctic Archipelago (see front map) to the north of Canada. Cnoyen's summary names 78° north latitude as the site of the indrawing channels between the islands. The Canadian islands happen to be centered on just this latitude. The primary islands also happen to number about eighteen or twenty, as shown by Clavus, Behaim, Ruysch, and Mercator. Clavus's latitude placement of the islands would be closer to the truth than the others' polar location. However, there are some other matters that demand attention if such an interpretation is to be upheld. Because the determination of longitude is more difficult than that of latitude, the east-west placement of the islands on Behaim's globe is questionable. It is more likely the result of the sequence of visiting described in the *Inventio* author's itinerary than of actual measurements. Behaim introduced a still further potential distortion in a graduated scale running from the North Pole southward past the left edge of Greenland. This scale is graduated not in degrees of latitude but in the length of continuous daylight at the summer solstice. Any actual information of the length of midsummer daylight on the Canadian islands would have been incompatible with the latitudes at which the islands are set down on the globe. Their days are much shorter than Behaim's location of them would indicate. They, as well as other known lands, such as Britain, should properly be placed much farther south, even on Behaim's own scale.

There is a reasonable explanation of why Behaim was able to accept the excessive northerly placement of the islands, probably even more northerly than the latitudes in the *Inventio* would have indicated. Namely, the length of a degree accepted by Behaim was considerably shorter than the correct value. Thus, the known lands of the earth covered a much greater percentage of his globe than they should have. This discrepancy was hidden in the east-west dimension by an excessive narrowing of the (then uncertain) distance westward across the ocean from Spain to the Orient. In the north-south dimension, however, the known lands show no compression to take up the discrepancy. Evidently the north-south compression all took place in the unknown polar region. Behaim was thereby forced to accept any northerly discrepancy with the latitudes in the *Inventio* in order to accommodate the islands at all. In summary, one need not be surprised that the Canadian Arctic

Archipelago is a likely candidate for Cnoyen's islands, in spite of their placement on the globe adjacent to the North Pole.

Cnoyen's description of the relationship between the nineteen channels and eighteen or twenty islands is far more detailed than the description of the islands themselves. An examination of this description may shed some light on the identities of the individual islands. The complete text of the description has been analyzed and translated into English by E. G. R. Taylor in her cited work.[8] I quote small sections of her translation in the following analysis. (Refer to Figure 5.) "In 'North Norway,' which is called Dusky Norway, there are three months of darkness during which there is no sunlight but a perpetual twilight." If "Dusky Norway" is a real place, it must, by astronomical reckoning, be either northernmost Greenland or Ellesmere Island.[*] The only other places approaching three months of darkness are Northland north of Siberia and Spitzbergen, and these do not make sense in connection with the rest of the description. Northernmost Norway itself has less than two months of darkness, and is ruled out anyway by Cnoyen's later statement that "North Norway" is one of the Septentrional Islands. As a starting point I will assume that "North" or "Dusky" Norway is Ellesmere Island, since Greenland would probably have been named explicitly.

"This North Norway lies over against the country called the Province of Darkness: in Latin *Provincia Tenebrosa*. [This would be Axel Heiberg Island, just true west of Ellesmere Island.] . . .And between this Province and Dusky Norway there is only 12 miles of sea [Nansen Sound and Eureka Sound]. From North Norway you cannot reach the Indrawing Sea, which lies beyond Grocland. For it lies still farther northward. This North Norway stretches as far as the mountain range which encompasses the North Pole, and borders on this mountain range for about 17 French miles by land: the rest is all sea. And this is the same mountain range which [text damaged here] . . . within about 15 French miles and then stands further off towards the east." This may refer to the fact that the lower part of Ellesmere Island is connected to the upper part by an isthmus about 17 miles wide. There is much uncertainty on our part between French miles and German miles, which were four times longer, because both are used in the narrative. Ruysch says that a magnetic rock at the pole was 33 German miles in circumference and Mercator says 33 French miles. The sentence with the missing text might have clarified this if it had not been obliterated by fire damage.

---

[*]Estimations of latitude such as this are presumably associated with original observations, not Behaim's scales, and should be given weight.

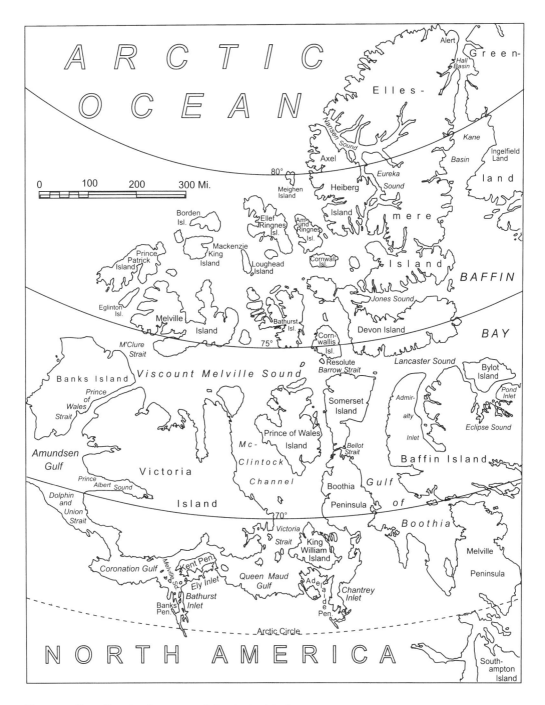

Figure 5. Canadian Arctic coast and Arctic Archipelago

"And near here, towards the North, these Little People [the Thules] live of whom there is also mention in the *Gestae Arthuri*. And there borders on it besides a beautiful open land. And this land lies between the Province of Darkness and the Province of Bergi." The Province of Bergi* would seem to be Devon Island and the other open land the smaller Ringnes islands between Devon and Axel Heiberg. Numerous Thule remains have indeed been found in this area, often in association with Norse artifacts, and Amund Ringnes Island is quite low and open compared to the other mountainous, glaciered islands.

Cnoyen's rendition of the *Inventio Fortunatae* and *Gestae Arthuri* must be very cursory, for I have quoted the most detailed part. Nevertheless, he has permitted us tentatively to identify the islands which in fact would have been the most readily accessible from Greenland. Beyond the islands so far discussed lie the "Indrawing Seas." Unfortunately, Cnoyen's narrative becomes so highly summarized at this point that it is impossible to interpret the details of the channels with any specificity. They could be anywhere. He does clearly indicate the farthest point reached. He says, "these innermost seas number four: and the one which lies on the west side was quite 34 French miles broad. And on the other side of this sea was the best and healthiest land in all the North." The western channel referred to could be construed as the McClintock Channel and then the "best and healthiest land in all the North" would be identified with Victoria Island (see Figure 5). In fact Victoria Island *is* the least hostile island in the entire Arctic Archipelago and comes within 120 miles of the tree line of the Mackenzie Forest; the midsummer mean temperature of afternoon high and nighttime low in Victoria approaches 50° Fahrenheit.[9] In the optimal climatic period of the Middle Ages, before onset of the Little Ice Age, it may have been warmer still.

The most prominent island Behaim depicted is the large westernmost island—in Victoria's correct place—shown, on his perhaps exaggerated scale, even larger than Greenland. It is also the only island containing a caption, evidently having impressed Behaim as being of some importance. Unfortunately, the Behaim globe is highly deteriorated in this area, and the caption is illegible. Fortunately, on the other hand, facsimiles of the globe were made when it was in a lesser state of deterioration. A planar facsimile made in 1730 interprets the caption as "*hie findt man weises volk*" (white people are found

---

*Cnoyen's (or the *Gestae Arthuri*'s) use of names from Marco Polo's *Travels* is appropriate under the one-ocean paradigm. However, we today need not accept Polo's Asian locale as a foregone conclusion in this reinterpretation.

here). This interpretation proved unsupportable, and a later facsimile in 1778 reinterpreted the caption as "*hie fächt man weisen valcken*" (white falcons are caught here). The latter interpretation has been accepted by most scholars ever since and is backed up by the appearance of similar captions on other maps. These captions have usually been assumed to come from Marco Polo's reference to falcons in northeast Asia. However, Marco Polo never referred to *white* falcons. Pure white gyrfalcons, unique to the American Arctic, were supplied to medieval Europe by the Norsemen. Remember that northeast Asia is equivalent to Arctic America under the one-ocean paradigm. Like Eskimos being seen as Hyperboreans, the presence of American white falcons would satisfy Polo's story, and the importance of falcons to ancient sportsmen is legendary. But Behaim's apparent identification of Victoria in preference to Greenland as a land of falcons seems perplexing. White falcons are found throughout the Arctic. A natural question would be, "Does Victoria Island harbor an unusually large number of falcons?" There have been two major expeditions to the area in modern times, but the Canadian Arctic Expedition has never published its report on ornithology and the ornithologist on the Thule Expedition (Feuchen) did not include Victoria Island in his itinerary. Some modern observers report the bird as rare on the southeast coast of Victoria (where there is a scarcity of the bird's preferred cliffside nesting sites), but others report nesting sites at various parts of the island.[10] The arctic falcon preys largely on ptarmigan, of which there are certainly more on Victoria (and throughout the Arctic) than on Greenland. We shall later see more evidence that Behaim's globe depicts Victoria Island.

This interpretive picture of the Septentrional Islands and Indrawing Seas fits many aspects of the real geography of Canada, but there is one aspect of the description that seems wrong. The seas and channels surrounding the Arctic Archipelago are *not* indrawing from the perspective of a visitor from Greenland. The average current in this area is eastward and southeastward, which is the reason these channels are almost always filled with ice from the Beaufort Sea and Arctic Ocean.[11] Nevertheless, there are specific real channels that could have occasioned the description "indrawing." Cardigan Strait and Hell Gate, between Ellesmere Island and Devon Island, form passages on the order of 5 miles wide connecting major bodies of water. Tidal flows there could well have been indrawing to Norse sailing ships. A similar situation obtains in Bellot Strait between Somerset Island and the mainland at Boothia Peninsula. Lancaster Sound and its connecting Barrow Strait are the only relatively safe entries to the archipelago.

I shall make one more observation concerning the story of the Indrawing

Seas. Cnoyen does give one description that is definitely pole-symmetric. He states that right at the pole there is a huge funnel-like whirlpool into which all the Indrawing Seas ultimately pour, surrounding a magnetic mountain. He states that it is 8 degrees of latitude in diameter, which amounts to nearly 500 land miles by most measures of the time. A true whirlpool of that size would be so energetic as to affect the weather of the entire Northern Hemisphere and would probably be unapproachable by humans or other living things. However, the sea into which the archipelago's channels terminate in the west, the Beaufort Sea, is famous for the Beaufort Gyre. This is nearly 500 miles wide. It is a stately circulation of grinding ice actually caused by the rotation of the Earth and prevailing winds. Its center could only be imagined by an observer at the edge, and it constitutes an invitation for speculation based on myths. An Eskimo observing its edge (or the *Inventio* author doing so) would not even perceive it as circulating but as moving along in a fixed direction. At the western edge of the archipelago it moves southward, shifting gradually to westward along the coast of Alaska. If the scientifically trained author (or some intermediary) pieced together several different Eskimos' stories from different vantage points, he could have perceived the arc of a circle and calculated an overall assumed diameter.* He could then, surmisedly but quite logically, center it on the North Pole. And at such removal and with such need for interpretation comes the opportunity for the earlier mentioned paradigm of identifying newly encountered real phenomena in mythical terms. Many compass users presumed a magnetic mountain at the North Pole, and other mythical material was potentially accessible to this scholarly author.[12] Another myth likely to have had influence here is the Norse story of Ginnungagap (see page 99).

One final item of polar symmetry remains in Cnoyen's words. He refers to a ring of mountains that circles the Earth at 78° north latitude. This probably arises from the Riphean Mountains, the great stone girdle believed by the Russ people to encircle the whole Earth.

Can one conclude that the Septentrional Isles and Indrawing Channels had some basis in reality and were not totally imagined experiences of the Norsemen? Perhaps not unconditionally, but it no longer seems unreasonable to study this possibility seriously. An origin in reality may have been

---

*In private communications, I have put the following question to two independent experts on practical Arctic logistics who have spent their lives working closely with Eskimos, Laurie Dexter and Arthur Mortvedt. "Could the Thules Eskimos themselves, with their moving about and sharing of geographical information, as a people, have perceived the circularity of gyration?" They both agreed that yes, the Eskimos themselves likely would have.

shaped by mythology. In this case the outlook at the source of these descriptions is definitely European, not Eskimo. While the Eskimos did give place names to localities, they seldom named land masses or water bodies, and to them the coherence of an island was of lesser importance than to us.

THE CIRCUMPOLAR AREA OF BEHAIM'S GLOBE (Figure 48) contains what he presumed to be the northern edge of the Eurasian continent. It provides considerable detail, not just the "edge of the world" found on most other maps of the time. Let us examine Behaim's depiction of northernmost Asia. The most striking feature is the great inland "Frozen Septentrional Sea," a feature completely foreign to any reality inside Asia.[*] As previously shown, the Behaim globe seems to contain information about the American North, and there indeed is an inland sea in North America, Hudson Bay. Behaim does not show it in the right shape or in exactly the correct relative location within the continent and does not show any connecting strait, but its very presence and size are suggestive. The Ruysch map (Figure 54) shows a similar inland sea of almost identical shape labeled *Mare Sugenum,* "Indrawing Sea." This name was used by Cnoyen in the singular at one place referring to a sea "beyond Grocland" and otherwise in the plural to refer to the channels between the Septentrional Isles. Hudson Bay does have certain indrawing tidal characteristics, and Hudson Strait has persistent indrawing tidal flows of five to seven knots, especially along its northern edge.[13] Furthermore, Hudson Bay itself, in conjunction with the Earth's rotation, supports strong tidal currents circling its entire shoreline counterclockwise.[14] If we can find on Behaim's globe any other North American features in this putative vicinity, Hudson's Bay's identification will be strengthened.

Scandinavia and the northwest part of Behaim's Arctic coast are based on his predecessors' interpretations of Claudius Clavus, which will be studied later. But there is no known precedent in the history of cartography for Behaim's central Arctic coast. A clue to a real-world origin for this graphic may be provided by the position of the aforementioned hypothetical Victoria Island with the white falcon inscription. The hammerhead peninsula on the coast almost touching this putative Victoria Island can easily be imagined as Kent Peninsula (see Figure 5). This skirts the southern shore of Victoria Island, flanked on the west by Coronation Gulf and on the east by Queen Maud Gulf. Proceeding eastward along the analogue of Queen Maud Gulf

---

[*]Ravenstein's conjecture that this represents the Arctic Ocean is a speculation without any supporting evidence, even that they knew there *was* an Arctic Ocean.

on the Behaim globe one comes to a second bulbous peninsula, which corresponds to the real Adelaide Peninsula abutting King William Island. Just east of this peninsula on the globe is an inlet which would correspond to Chantrey Inlet. Beyond that to the east, the coastline becomes smooth and sketchy, and the recognizable Canadian accordance seems to end.

Returning to the first peninsula and proceeding westward, the next feature of note is a highly convoluted inlet, essentially a deep inlet with an interior hammerhead peninsula. Except for one flaw, that inlet is of the correct character and location to be identified as a combination of Melville Sound and Bathurst Inlet with its interior Banks Peninsula. The flaw is that, on a scale appropriate for the other identifications, this Bathurst Inlet analogue would be too large and separated too far from Kent Peninsula. Nevertheless, local scale disparities in Eskimo maps are well known (and apparent in Figure 1). If this entire graphic from the base of the Kent analogue to the left end of the Melville Sound analogue were reduced by a factor of three, it would become commensurate with the other identifications, and the stretch of coastline separating it excessively from Kent Peninsula would become the inner coastline of Elu Inlet.

This rationalization naturally takes something away from any confidence we might have in such identifications. Furthermore, these correspondences are not as comfortable in their individual details as those in Clavus. Nevertheless, their occurrence in the appropriate sequence should be weighed in deciding if they are forced or accidental. The multitude of appropriate details here certainly describes an extended coastline rather than isolated minor wiggles. Other maps to be studied later will also correspond to this part of the Canadian Arctic coast.

CHAPTER 4

THE YALE VINLAND MAP

SINCE THE YALE VINLAND MAP (Figure 36) first entered scholarly study in 1957, its authenticity has been controversial. There have come to be so many theories of forgery of this map, many of them mutually exclusive, that it would seem that none of them is uniquely convincing enough to quell the birth of newer theories. Nevertheless, the majority of scholars seem to accept on faith that it must be a forgery. The underlying reason for this is the occurrence of the island of Greenland and the island labeled "Vinilanda Insula." It is simply incredible that this island and the adjacent inscription describing Vinland's eleventh-century discovery would have appeared on a map from 1440 without the appearance of similar information on any intervening map. The map seems too sensational to be anything other than a forgery. However, there is a way to interpret it in much less sensational terms, as simply another misunderstanding springing from another Thule Eskimo encounter contemporary with the cartographer.

Before demonstrating this interpretation, it is appropriate to examine several of the major forgery theories. The controversy almost ended in 1974 when a study by Walter C. McCrone Laboratories announced that the map's ink contained a pigment with twentieth-century characteristics.[1] The study concluded that this was absolute proof that the map was a forgery. But many people intimately familiar with the complicated details of the map's first emergence felt that no forger could have orchestrated the required coincidences.[2] A new ink study announced in 1987 by researchers at the Crocker Laboratory showed that the 1974 study had overstated its conclusions and that the offending pigment was present only in trace amounts.[3]

I had meanwhile shown a logical fallacy in Walter McCrone's reasoning: he had made an unstated assumption that the ink had to have contained the modern pigment since its origin. I showed in 1977 how traces of such pigments in the map's ink could have been introduced by a misguided cleaning attempt of a former modern owner.[4] The pigments could have been transferred into and throughout the ink accidentally and unknowingly during a traditional routine bleaching. Details of this transfer are described in the Appendix. This possibility renders the formerly absolute ink anachronism now one of several circumstantial curiosities.

Other such circumstantial curiosities are marshaled by Kirsten Seaver in her theory that the Vinland Map was forged during World War II as an intellectual trap for Nazis promoting Nordic accomplishments.[5] The speculations supporting this theory are tempting primarily to scholars who already reject the map's authenticity. Without analyzing supporting detail here, Seaver's Nordic pride theory is built around essentially two interesting pieces of evidence: (1) The Vinland Map's telescoping of Leif's and Bjarni's voyages into a single one parallels a similar telescoping in a 1765 publication; (2) a sale catalogue described a *Speculum Historiale* fragment that could have fit a missing portion in Yale University's *Speculum,* which accompanied the map. However, the telescoping of unfamiliar information is a widely known human trait and need not be made overly significant. Also, the *Speculum* was a widely copied document, many fragments of which are still circulating; without the actual fragment to examine, this need not be made overly significant. Both of these points may still turn out to be will-o-the-wisp coincidences more appropriate to historical novels than historical research.

Meanwhile in 1996 the Crocker group showed how nonrandom selection of particles for the analysis by the McCrone group could have biased their results.[6] Indeed, Walter McCrone has admitted on recorded television that even before any samples were taken, "My mind was made up when I first saw the yellow ink line itself."[7] Moreover, Ardell Abrahamson has shown that the presence of modern-seeming pigment in the ink could have occurred in nature, leaving the anachronism argument irrelevant. Furthermore, internal evidence, including reinstatement of once discredited cryptograms discovered in the map's inscriptions, continues to accumulate and to make a case for its possible authenticity. In 1967 cryptographer Alf Mongé discovered a hidden message in one of the map's inscriptions; it followed medieval rules he had rediscovered only after the map's emergence, rules therefore unavailable to any forger. Critical reception of Mongé's cryptographic arguments was negative, partly because of reviewers' confusion of the Vinland Map cryptogram with other cryptographic fallacies that exploited large degrees of freedom. But the map's cryptography, as analyzed in my own 1991 publication, has much less of what statisticians call "wiggle room," and it remains unrefuted.[8] Cryptograms were to medieval scribes what computer hacking is in today's world. The vast majority of early modern hackers intended no harm, looking merely to display their hidden cleverness to a restricted audience capable of appreciating it. Medieval scribes themselves formed such an audience for medieval cryptograms. For a modern forger to expend the tremendous creative energy needed to make these cryptograms with no likely appreciative audience would be pointless.

In 1998 Paul Saenger produced a long list of suspect nits in the circumstances of the map's emergence into modern study.[9] To someone predisposed to a conspiracy theory, they are all highly damning. However, they are all highly subjective. While they deserve study by anyone willing to pursue them, they might each conceivably be explained by some circumstance other than forgery (such as sharp business practices). Saenger also makes the complaint voiced by many others that Yale University has not actively supported any research by recognized scholars that might cast further light on the map's authenticity or lack thereof. This fact is explained by an explicit and announced decision by Yale after the McCrone debacle that they would no longer involve the university formally in further study of the map, leaving this to individual scholars at their own will.

As long as there is no absolute proof in either direction, there is every reason to entertain a possibility that the map is false; but at the same time, one should be open at least to looking at the circumstantial evidence for its truth. Yale University Press's republication of the original 1965 map study in 1995 included a review of evidence supporting its authenticity.[*][10] Since that case will be further strengthened by other evidence herein, I will proceed to take the map seriously, even while acknowledging that authenticity is still not (and probably could never be) absolutely proven.[†] It is possible to be agnostic and still make a serious study of religion, and such is my attitude concerning the Vinland Map.

THE YALE VINLAND MAP (as opposed to other known Vinland maps) is a world map in the relatively unusual oval disc school of cosmology,[11] hypo-

---

[*]The press did not fund the research, and it is in any case a separately self-sustaining entity within the university.

[†]In fact, no artifact can ever be absolutely proven authentic, only inauthentic. To prove inauthenticity, one need only demonstrate overwhelming evidence of fraud; but to prove authenticity would be to prove the impossibility of any such evidence ever, a logical negative proof, which has been recognized as impossible ever since Plato.

thetically drawn about 1440 by a southern European author. It depicts, besides the generally known medieval world, the island of Greenland and an island labeled "Vinilanda Insula." The island so labeled bears no apparent resemblance to any known coastline.* The depiction of the Greenland coast is strikingly accurate, however, and is one of the most arresting features of the map. So arresting is the accuracy of the Greenland coast that it has apparently diverted attention from a significant distortion of Greenland. The true length of Greenland is roughly three-quarters the length of the Mediterranean Sea. (Although medieval maps often exaggerated the Mediterranean's length, it is still a useful rough scale reference.) But on the Yale Vinland Map Greenland has been drawn at about half that proper relative size. I will speculate on possible reasons for this distortion later in the chapter.

Figure 6 shows how the Yale Vinland Map would look if Greenland were expanded to the proper scale. While expanding the scale of Greenland I have held the scale of the unknown land named Vinilanda fixed and kept the location of the northeast corner of Greenland fixed. (Any assumption of correct latitude is no longer sustainable in this case.) The immediate visual effect of this is to make the formerly imposing Vinilanda Insula seem much less important. The second effect is to remind one of the unknown island labeled Grocland west of Greenland in Gerard Mercator's map of 1569 (Figure 68b). While Vinilanda Insula is still too large compared to Grocland, and the orientation is off by nearly 90 degrees, the similarity of shape and location is nevertheless striking. Note the concavity in the coast away from Greenland in both. Mercator (i.e. his source Cnoyen) states that the Island of Grocland was settled by Scandinavians, bringing to mind the Yale Vinland Map's inscription about Vinilanda Insula's discovery by Scandinavians.

The cartographic resonance between Grocland and Vinilanda is reinforced by a semantic resonance. At the turn of the twentieth century Sven Söderberg proposed that in the Norse word *Vinland* the root *vin* with a short *i* originally referred to "pasture" rather than "grapes." In the 1980s Gwyn Jones and Eric Wahlgren argued against the Söderberg thesis; Wahlgren called it a "will o' the wisp."[12] Einar Haugen has objected that that usage of *vin* had disappeared in the written record by the time of Vinland's discovery.[13] However, those arguments against Vinland as pasture land are based strictly on negative evidence. There is an equal lack of evidence that wine had any importance

---

*Some have assumed correctness of latitude in this location and postulated that it is a schematic rendition of the northeastern quarter of North America, including Hudson Bay and the St. Lawrence estuary, the western coastline being an incorrect surmise.

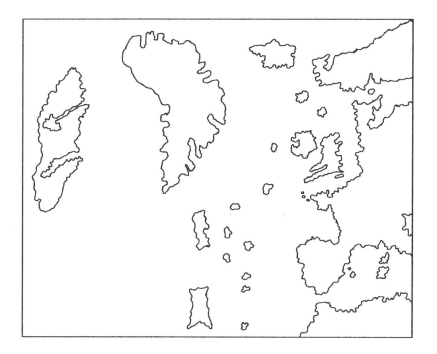

Figure 6. North Atlantic portion of Yale Vinland Map as it would have appeared with Greenland drawn at correct relative scale

to the original Greenlanders. Making wine from wild grapes requires the addition of sugar for fermentation, or wild grapes yield an ignoble rot instead of potable wine. The only source of sugar in the North was the precious honey that was used instead for mead. Suggestions that wine was made from berries fail to acknowledge that berry winemaking also requires sugar. Haugen meanwhile does acknowledge the existence of some positive evidence for Vinland as pastureland, or at least farm fields. Helge Ingstad has strongly supported this interpretation.[14] My 1972 book gave detailed positive arguments for the pastureland semantics that nobody has specifically refuted.[15]

The point relevant to the semantic resonance between Grocland and Vinland is this: Cnoyen's "Belgic" root *groc* falls very nicely between the middle Dutch *crocke,* the Flemish *krakke,* and the modern Dutch *krok,* all with approximately the same meaning—wild pasture. Apparently Söderberg was not the first to adopt the *Vinland* as "pastureland" interpretation. Thus, we have a semantic as well as cartographic resonance between Vinland and Grocland.

Consideration of all these facts leads to the following bold hypothesis: that the Yale map's Vinilanda and Mercator's Grocland both represent Baffin Island, just west of Greenland (see front map). The Thule Eskimos did in fact migrate through Baffin Island. Furthermore, a recent discovery of Norse yarn in a Baffin Island Dorset Eskimo site suggests some kind of interaction there.[16] As a check of scale, note that the real length of Baffin Island is about the same as the distance from Gibraltar to Ireland, just as with Vinilanda Insula. Recall that Cnoyen stated that the indrawing sea, which Ruysch described in the singular, could only be entered by going beyond Grocland. This is exactly the situation between Hudson Bay and Baffin Island. So the two independent interpretations—Grocland as Baffin Island and the singular Indrawing Sea as Hudson Bay—support one another.

Momentarily, return to the Martin Behaim globe (Figure 48) and recall the island with an archer and bear that we assumed was Greenland. Just west of this Greenland, nestled in the gulf of its west coast, lies an island corresponding to Grocland. If this is rotated 90 degrees counterclockwise it gives as good a likeness of Baffin Island, as this Greenland is of Greenland.

This is not to suggest that Leif Erikson's Vinland was Baffin Island. The label *Vinilanda Insula* is a curious term. Vinland in the Sagas was never referred to as an island (although it admittedly had a minor nameless island lying off its north-pointing promontory). But the Sagas did refer to another island and only one, by name. That was Helluland. Perhaps the label Vinilanda Insula arose out of a confusion about Helluland. The great majority of scholars who have studied the Sagas have concluded that Helluland was Baffin Island.

The intuitive idea that the Yale map's Vinilanda might represent Baffin Island has been suggested by people ranging from Harvard University's Samuel Eliot Morison to a fourteen-year-old high school acquaintance of mine.[17] I will attempt to put some substance behind the intuition. The Yale Vinilanda shows inland water features as well as coastal outline. It is very unusual for legendary, semilegendary, or obscure islands on the periphery of medieval maps to show inland water features, and their presence suggests that this map does not fit those categories. The lower inlet (see Figure 6) might be identified with either Cumberland Sound or Frobisher Bay, shown on Figure 7. The southeastern coastline is quite like the southern coastline of Baffin Island along Hudson Strait, with Foxe Peninsula pointing off to the left. The concavity along the west coast is suggestive of Foxe Basin. The triangular-shaped upper bay resembles Eureka Sound behind Pond's Inlet. On the other hand, the dog-legged shape of the lower inlet is like neither Frobisher Bay nor Cumberland Sound, and there is only one such inlet. Also, the upper inlet is much too far inland, and the orientation of the features is not correct.

To address these anomalies, it is useful to have some understanding of how a map of an area as large as Baffin Island was made before aerial surveying.

No one individual, be he an Eskimo, a Norseman, or an Englishman with an astrolabe, is likely to have made a grand survey recording all details of the coastline. Rather, he would have collected local maps or descriptions already from local cogniscent individuals and pieced them together onto the preexisting background. This has been the main job of the cartographer throughout history (except perhaps in high medieval Europe, when tradition was supreme). The farther the cartographer is removed from knowledge of the real situation, the greater the likelihood of error, particularly errors concerning adjacency and orientation of each geographic piece.

Let us entertain the candidacy of Baffin Island and attempt to create Vinlanda Insula from it. Figure 7 incorporates within its coastline Bylot Island above Pond Inlet. Figure 8 shows one compilation of eight regional maps representing hypothetical local knowledge. Eskimo maps generally concerned themselves with such local knowledge, while European cartographers preoccupied themselves with the "big picture." Map 1 covers Foxe Peninsula and the Hudson Strait shore; map 2, Frobisher Bay; map 3, Cumberland Sound; map 4, Cumberland Peninsula and Home Bay; map 5, the northeast coast; map 6, Eclipse Sound behind Pond Inlet; map 7, the northern reaches; map 8, Foxe Basin. Franz Boas collected some Eskimo maps that correspond to regions 2 and 3.[18]

In Figure 9 these local maps have been reassembled by a hypothetical European cartographer, incorporating the following errors: (1) Regional map 4 has been "lost"; and (2) Waterway regions 2 and 6 have been *attached to the inner ends* of the waterways in maps 3 and 7. I am imagining the activities of a cartographer sitting somewhere in Europe, not the Norse explorers or Eskimos themselves. They were presumably quite familiar with the actual situation and had no trouble getting about. It is unlikely that any Norseman or Eskimo ever saw the Yale Vinland Map.

I am certainly not proposing that this admittedly far-fetched scenario is exactly how the Yale Vinland Map was produced. I manipulated some parameters in constructing this figure but required nothing like E. G. R. Taylor's machinations attempting to prove the map a forgery.[19] I am illustrating the kind of things that can happen with separate data sets and incomplete and misunderstood information. There is another point to be made about this scenario. Note that on the front map, the two lower bays of Baffin Island point approximately toward the southern tip of Greenland. This kind of information might have been of significance to explorers out of southern Greenland and become associated with the regional data sets. If the cartographer of the Yale Vinland Map had no other information than these ex-

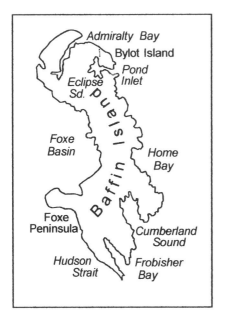

Figure 7. Outline map of Baffin Island, the western neighbor of Greenland

plorers had, this could explain why Greenland was drawn at half scale on the Yale map. In it, the rotated bays do point to Greenland's southern tip, which they could not do at full scale.

Attention to these waterways leads to a further possible cartographic resonance. Whatever was the cartographic process of putting the bays into Vinilanda Insula, it nearly cut the island apart. There is another map in which such cutting appears to be complete: the 1507/8 map by Johann Ruysch (Figure 54). Note the two small rather artificial-looking islands of semicircular shape that he places symmetrically about a central channel just west of Greenland, in what he calls the "Sinus Gruenlanteus," that is, "Gulf of Greenland." Ruysch and Mercator are known definitely to have had a source in common for this part of the world, the aforementioned *Inventio Fortunatae* or Cnoyen's *Itinerarium*. Note that Mercator's rendition of Grocland is, like Vinilanda, quite naturalistic in contrast to his Septentrional Islands. Even though Mercator's extract of Cnoyen in John Dee mentions Grocland, the Dee text which we know does not seem to preserve the complete source of Mercator's Grocland graphic. Mercator's letter to Dee was in response to Dee's specific questioning about the Septentrional Isles surrounding the pole and their connection with King Arthur. Dee may well have omitted a part of Cnoyen's narrative that included more descriptive detail about Grocland, perhaps even cartographic detail, irrelevant to Arthur. Indeed, Dee's transcript is preceded by this comment regarding Mercator: "This then with other matter, I receyved from him lately." The bifurcated island depicted by

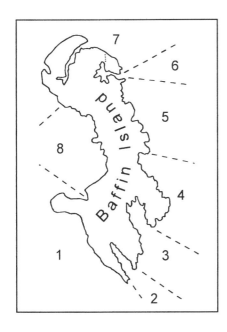

Figure 8. One compilation of eight hypothetical regional maps of major features of Baffin Island

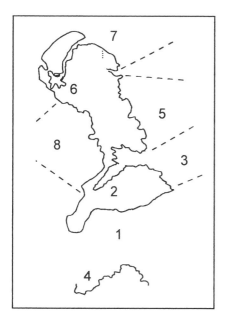

Figure 9. Incorrect arrangement of accurate regional maps, to simulate Vinilanda Insula

Ruysch, although he does not give it a name, had been shown in the same form by other mapmakers somewhat before his time. The other mapmakers always identified this strange island as an unknown "Island of Brasil," the Irish "Fortunate Isle," sought in late medieval times. Written traditions spoke of the island as having a river running through its middle. Perhaps at some subtle level the Yale Vinland Map is also connected to the *Inventio Fortunatae*. The

Inventio may have contained some descriptive data about Grocland that could, depending on how it was interpreted and passed on, have yielded either Mercator's Grocland, the Yale map's Vinilanda Insula and waterways, Ruysch's channelized island, or the Isle of Brasil. Note that the first three of these have always been associated closely with Greenland. A 1480 Catalan map in Milan not shown here depicts Brasil associated with Greenland, as do some other fifteenth century maps.[20]

The conclusion that Vinilanda Insula could represent Baffin Island does have some bearing on the perception of the Yale Vinland Map's authenticity. Scholars' initial assumption was that Vinilanda Insula represented the Canadian seaboard, because they imagined suggestions of Hudson Bay, the Labrador coast, and the St. Lawrence River and Gulf. This forced them to dismiss the rear coastline as a (wrong) surmise by the cartographer. That assumption of mainland was based partly on the apparent correspondence of latitude of the features where the cartographer had placed the island. But it was also based in large part on wish fulfillment, especially the cavalier dismissal of the rear coastline defining its insularity. The Yale Vinland Map was coming into scholarly study contemporaneously with the discovery of Norse ruins in Newfoundland, and there may have been a natural psychological attraction to closure. For conservative thinkers, the enthusiasm about the new discoveries was a red flag calling for skepticism. Skeptics devoured points against the map's authenticity with as much enthusiasm as the pro camp did points in its favor.

The likelihood that Vinilanda Insula merely represents Baffin Island lowers that red flag to some extent. Obviously it cannot change the fact of authenticity either way, but it can change the psychological expectations about authenticity. It is difficult to imagine a forger's creating an artifact that fits so well into the historical record and yielded no gain for the forger. One slightly tenable theory of nonauthenticity has been that the map is a mind-game created by an early-twentieth-century scholar to amuse himself.[21] One suspects that, in such a case, he would have picked something more exciting than Baffin Island; or, if he had discovered the foregoing material himself, he would have published it.

As we proceed, we will see more and more that the Yale Vinland Map fits into a chronology that makes us quite open to its possible authenticity rather than to rejecting it forthwith. If this map were in fact authentic, it would complete a scenario of Eskimo information of Arctic America from one coast to the other and reaching to southern Europe.

II

The
Chronological Survey

CHAPTER 5

INTRODUCTION TO THE

CHRONOLOGICAL SURVEY

WE HAVE NOW EXAMINED SOME OF THE MORE STRIKING cartographic evidence for the idea that, early in the Renaissance, Eskimos and Greenlanders might have been involved in the European discovery of America. But to gain real confidence in our hypothesis, it must be subjected to an exhaustive set of tests. For this idea to merit the higher appellation of "theory" or "conclusion," it must demonstrate two qualities. It must be consistent with all other known aspects of history, and it must be capable of extending the present understanding of history. A meaningful start can be made using one common technique. That is the chronological survey, in which all relevant history is reexamined in summary form with the hypothesis in mind at each point.

The remainder of this book contains a chronological survey. Its format encompasses a body of information whose meaningful presentation requires acute attention to detail, as the study of maps does, combined with the insights attainable from a general overview. W. T. Barry and H. E. Poole produced an entire history of printing this way. In the field of the history of discovery, Henry Harrisse has shown that a chronological parade of cartography and voyages not only provides a framework for otherwise unattainable insights but can also make good reading.[1] I hope to present something not far short of an exhaustive list of known information relating to the medieval concepts of Scandinavia and the North (including the West). There is a sequential directory to the Chronological Survey in the front matter of this book. By such an exhaustive survey we hope to complete the hypothetico-deductive model suggested in the Introduction. My hypothesis is that divulgence-hiding paradigms can explain *every* instance of apparently fantastical, otherwise unexplained shapes in the Arctic and Far East on maps made after the Thule-Norse encounter. I fully expect challenges to this statement, and that is how progress is made, but we will see that the statement is more than just a sacrificial lamb. Upholding such a hypothesis could be one way to counter an otherwise potentially valid accusation that we are simply marshaling isolated evidence to suit a foregone conclusion.

The Chronological Survey also provides a supportive framework for examining the plausibility of interpretations that by themselves might never be considered. In the process of this survey I shall diverge from the standard way historians have looked at many documents. The survey will occasionally repeat some information already presented, when its appearance in chronological context is important.

The subject of medieval Scandinavian geographical awareness has been full of puzzles for researchers. It is difficult to find any potential entry for the list which does not contain at least one of these puzzles. In the course of the chronology threads of analysis will be formed which are not apparently related to the problems posed so far. These threads do converge on solutions to the puzzles related to the western explorations, sometimes unexpectedly related.

In the Middle Ages, as now, maps were often discarded when a new edition came along. This has left us with a spotty record to study. And the history of ancient manuscript collections is also the history of fires, floods, and vandalism.[2] Manuscripts and documents that remain in existence form isolated windows into the past. Sometimes a light shone into one illuminates the view from another. The dates used here for chronological ordering of the material being discussed have been settled upon by many scholars, and I will consider them as relatively established even if a few remain controversial. Thus, the time relationships of events and authors will be considered important, evidentially, as will the contents of the manuscripts, maps, etc. Mapmakers have often copied information about given areas from their predecessors, forming entire graphic lineages. In these cases I will consider only the earliest known in a line containing specific new information. Later variations might be attributable to unknowable influences, like cartographers' individuality, and would not make good subjects for our kind of study.

For each item discussed in this survey, a separate list of bibliographical references is given in an endnote. These lists are not intended to be either complete or representative, but thorough bibliographies may usually be found in the works cited. Some of the reproductions of maps used in this book are processed reconstructions, which bring a clarity to elements of the original maps, frequently unclear or faded.

The ability to understand medieval maps is sufficiently important in this reading that a few sentences on their interpretation will be useful. Early and high medieval cartographers made no use of the scientific observations that were so important both before and after their era. For their purposes, symmetry and authority were more esteemed than accuracy and truth.[3] The purpose of their wheel maps, or mappemondes, was not to show reality for a traveler. It was to rationalize and justify their literal interpretation of statements in the Old Testament and to provide an anchor for Christian symbolism.[4] People in the High Middle Ages tended to forget about the geometric origin in theory of the disk-like Orbis Terrarum. Many laymen looking at disk maps did not visualize them as actually representing the dome of Orbis Terrarum on a spherical earth. That does not necessarily mean they thought the Earth was flat, just that they never thought about it at all. Those with a practical need, such as seafarers and cosmographers, continued to be aware of the Earth's sphericity within the disk maps.[5] But even cosmographers gradually drifted away from the maps' scientific origins, later acquiescing to the idea that Jerusalem defined the center.[6] This concept eventually led to some simplistic throw-back arguments that the Earth must be flat, like sixth-century B.C. Anaximander's disk-earth floating on the waters. This was true, the argument went, because otherwise the center of the universe would not be Jerusalem but a point buried and inaccessible at the earth's interior. Such dogmatic sophistry could be and often was countered by those who cared about reality.[7]

The maps in this collection have been preserved until now because they were used to illustrate medieval philosophical and scholarly works. Very few of any practical maps that supplemented oral or written directions for travelers have survived. Thus the lack of apparent scientific progress in medieval maps was not necessarily caused by any lack of increasing geographical exploration. As Michael Andrews says, the classification of the disk maps "will not disclose any constant improvement, either in contents or in execution, but rather debasement, deterioration and a growing tendency towards symbolic and diagrammatic treatment."[8] Some such comment might be made about the entirety of medieval rational thought.

Nevertheless, the makers of the maps were forced to retain some kind of contact with reality. In particular, the relative placement and orientation of the various countries and cities was always as correct as was possible. The four cardinal directions were always observed rigorously, but the arbitrary modern (as well as classical Greek) tradition of placing north at the top was not adhered to.[9] Some maps were rotated 180° from our familiar conception (see

Figures 26, 37, 39, 42). Many were rotated 90°, placing north at the left and south at the right, leaving east at the top (see Figures 12, 14, 17, 18, 20, 21, 22, 35). The generally circular shape of medieval maps usually leaves little hope of recognizing continents by the shape of their coastlines. A starting point for orienting oneself may be had in the Mediterranean Sea, which is always present (see for example Figure 17). The orientation may be recognized by looking for a large inland sea that connects with the ocean in the quarter defined as west (Gibraltar), and then looking for inscribed Latin names of familiar countries in their appropriate relative locations. Since most of our interest will be in the northwestern quadrant, we can quickly localize ourselves after finding Gibraltar. Next, look for islands labeled "Hibernia" (Ireland) and "Brittania" (Britain), then for the Scandinavian region by adjacency if not by inscribed names. Always consult the front map when in doubt.

Finally, a short survey of Norse navigation is in order at this point, particularly with regard to North-finding. Some old map inscriptions that we will later see carry messages in northern waters like "These waters are sailed without use of the compass" and other statements claiming that northerners navigated without compass or chart, carrying it "all in their heads." Norse navigation is indeed poorly understood, but it is clear that they did carry much in their heads. It appears that they memorized rhyming tables of daily solar declination. We know that the ancient Polynesian navigators carried star maps in their heads, so it is certainly conceivable that Norsemen did the same.[10]

The frequent fogs and overcast of the North could be expected to obliterate the sun and stars for at least part of any journey. Ramskou speculates that the Norsemen were aware of naturally occurring crystals that react to polarized light by changing color when aligned directly with the azimuth of a cloud-covered sun (the effect occurs only when there is considerable blue sky overhead). The use of crystals in navigation does not seem to have been widespread, however, and was viewed with wonder.[11] Writers, both contemporary and medieval, have speculated excitedly about the power of this technique without understanding how it works. Sunlight passing through clouds is *randomly* polarized, but sunlight reflected downward from air molecules in a blue-sky hole in the clouds is linearly polarized in the direction of the sun. Anyone with a polarizing filter on an SLR camera can experiment with this. The description given in St. Olav's Saga (*Flateyarbók,* col. 466) confuses solar elevation with azimuth, and its description of use during a snowstorm is surely hyperbole.

Another suggested means of north-finding, dubbed the "solar compass," has been aggressively promoted by Soeren Thirslund. Its supposed principle of operation has been inferred from markings on a fragment of a wooden disk found in the Greenland ruins. However, intensive discussion on several Internet scholarly discussion lists indicates that this is purely a figment of the modern imagination.[12] While the idea of a solar compass would not have been beyond the Norse capability, the specific wooden disk that constitutes its primary evidence turns out to be just as likely a fragment of a Norse gaming piece. There is no further evidence for the existence of a solar compass in the Middle Ages.

The history of the introduction of the magnetic compass in the West is still controversial, being sometimes ascribed to the Norsemen and sometimes to the Mediterraneans.[13] This controversy has obscured the fact that the Europeans and Norse actually acquired the compass in the same era. It attained general use on Italian ships in the 1200s.[14] However, as mentioned above, there is some question about whether the Norse immediately made use of the compass in navigation. The first mention of magnetic navigation in Norse literature is in Hauk Erlendsson's edition of the history of Iceland's settlement (*Hauksbook*), which can be dated only by the date of his death as 1334. Here Hauk casually mentions, in relating the search for the headlands of Iceland many centuries earlier, that those seamen did their navigating without use of any lodestone. Such a casual mention implies that he expected his armchair readers to have a casual familiarity with what this lodestone he was talking about was. It also implies that navigating without the lodestone seemed more remarkable in his day than navigating with it.

Naturally occurring lodestones are formed by magnetic induction when lightning strikes a magnetite deposit. The earliest applications consisted of placing a piece of lodestone in a floating wooden bowl and allowing it to rotate into alignment with the north magnetic pole. Better response was soon obtained by using just a sliver of lodestone placed on a floating wood chip. Finally, it was realized that the lodestone's essence could be transferred to an iron needle by stroking it, and this needle could be floated on a straw. But no one yet realized that there were two poles, geographic and magnetic. The magnetic pole is believed to be caused by dynamic events deep inside the Earth's core which are not perfectly synchronized with the geographic rotation. The variation of the compass caused by noncoincidence of the magnetic pole with the geographic pole (Figure 10) was generally not understood and was ignored in southern Europe even past Columbus's demonstration of its importance on long-distance voyages. As late as 1500, Colum-

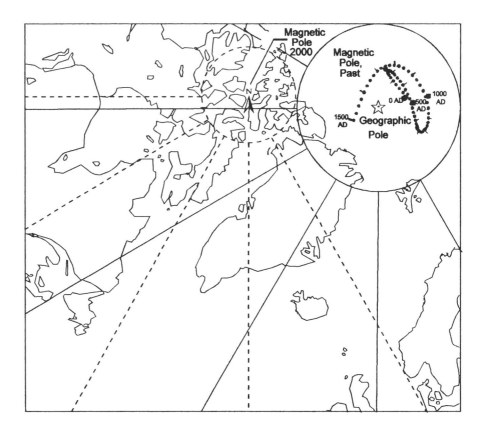

Figure 10. Geographic north compared with current magnetic north. Past north magnetic pole positions are plotted at 25-year points with ticks at 100-year points.

bus's own chartist, Juan de la Cosa, drew a compass chart of North American lands which made no allowance for this variation. Mercator, in 1569, was one of the first geographers to attempt to deal with the differentiation between the magnetic and geographic poles.

While compass variation is understood in detail today, its history has been poorly understood. The location of the magnetic pole itself has varied through history, and north and south magnetic poles switch position entirely at intervals of tens of thousands of years. It has recently become possible both to measure and to calculate their past positions, calibrated in geological deposits.[15] Figure 10 shows the most reliable part of such calculations in the whiplash path as the magnetic north pole has begun the next cycle of working its way southward.

In northern waters, variation today is quite severe. However, in the medieval period variation was considerably less than today, with the magnetic pole much closer to the geographic pole and certainly more in line with Europe. With the compass in use, the Norsemen's best policy would have been to switch mostly from celestial navigation to magnetic navigation—forget the stars and follow the stone. That was exactly the attitude taken by sailors of the Mediterranean in the period we are discussing.[16] The mystery of why those earlier sailors never discovered the variation may be answered simply by the fact that it was not yet severe enough for them to notice.

CLASSICAL NORSE GREENLAND

CHAPTER 6

EARLY SCANDINAVIAN GEOGRAPHY

MANY INTELLECTUAL CURRENTS, ranging from ancient Greek and Arab astronomy to Christian church history, contributed to the intellectual climate at the end of the first millennium. That climate was forced to assimilate the discovery of Greenland and Vinland. I will recreate here only the barest skeleton of that setting, in paragraph-length articles. The format, although it makes for somewhat choppy reading in this chapter, will justify itself in following chapters. I hope to reveal enough common ground within this intellectual diversity to contextualize Greenland's emergence and establish a thread that will run through later chapters.

A.D. 90–168: Claudius Ptolemy's Geographia. Any survey such as this must start with Ptolemy as a benchmark.[1] He lived at the height of the classical Greek development of the science of astronomy. In later life he published his now famous book *Geographia*. This introduced to a wider audience the idea of degrees of latitude and longitude on a spherical Earth and the means of constructing flat maps using these degrees (Figure 11). During the Dark Ages, such scientific principles were forgotten by most of the world but preserved by a few Greek and Arab scribes copying old books. Although the mapmaking art itself declined, some of Ptolemy's geographical conceptions, expressed in his map, persisted. The most important such conception for us was the erroneous idea that Scandinavia was a small island, unconnected to the Eurasian mainland, as shown on his map to the east of the Danish peninsula. We will see the general conception of Scandinavia evolve considerably as new discoveries are encountered.

Throughout Middle Ages: Legend of the Fortunate Isles. The legend of the For-

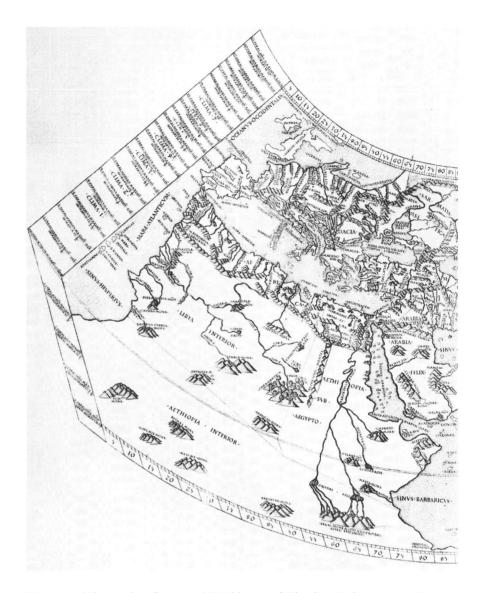

Figure 11. *(Above and on facing page)* World map of Claudius Ptolemy, A.D. 168 (copy, Rome 1478). Location: Multiple originals, including Göttingen, University Library. Source: Kamal, *Monumenta Cartographica,* vol. 5, fasc. 1, fol. 1498.

tunate Isles presumably originated with the desire to find some earthly paradise, and almost every discovery of new islands boosted it. The general idea extends as far back as the Elysian Fields of the Greeks. In later times Isidore of Seville summarized its description:

> The *Insulae Fortunatae* denote by their name that they produce all good things, as though fortunate and blessed with vegetation. For of their own nature they are

rich in valuable fruits. The mountain ridges are clothed with wild vines, and grain and vegetables grow wild like grass; thence comes the error of the heathen, and that profane poetry regarded them as Paradise. They lie in the ocean on the left side of Mauritania [south of Morocco] nearest to the setting sun, and they are divided from one another by the sea that lies between.[2]

It is easy for modern readers, aware of the finiteness of the globe, to over-look the importance that the idea of an earthly paradise held for the ancients and medievals. It must not be dismissed condescendingly. The only Paradise conceivable to an ancient layman was some piece of land far off in the sea, which the "infinite sea" separated from the known lands of human frailty. These ideas persisted through the Middle Ages, and only today has it become necessary to put Paradise completely off beyond space.

As Isidore states, it was nevertheless not proper to regard the Insulae Fortunatae as Paradise, but as some more intermediate state of blessedness. Certain real islands of the western ocean, notably the Canarics, were known in preclassical times by the Phoenicians and were still known in Roman times. Vague memories of these, which are indeed blessed by a moderate oceanic climate, probably interacted with the mythical material to produce a lively interest in their whereabouts. Ptolemy used the term *Insulae Fortunatae* to refer to the Canary Islands, his zero of longitude. The more important interest for our study comes after the discoveries of Iceland and Vinland.

Circa 874: Icelandic Settlement. Iceland, also known as Thule, was already inhabited when the Vikings found it in 860. Monks from Ireland had sought there a place to escape from worldly men and to pursue their studies and meditations. These holy men eventually went elsewhere, but they were an important part of the early history of Iceland. Some Icelandic place names still refer to the Irish. The name *Iceland* arose from a failed early Norse settler rueful about his inadequate winter preparations. The permanent settlement by Norsemen in 874 had timely political repercussions and brought new life to old legends about the world's remotest islands.[3]

885: Abu'l-Qasim Ibn Khordâdhbeh. Ibn Khordâdhbeh was one of the earliest Arab geographers to mention the Fortunate Isles of the western ocean (Atlantic). As the Caliph's postmaster in Medina, he wrote a "book of routes and provinces," which stated, "As concerns the sea that is behind the Slavs [i.e., to the north and west of them], and whereon the town of Tulia lies, no ship travels upon it, nor any boat, nor does anything come from thence.[*] In like manner none travels upon the sea wherein lie the Fortunate Isles, and from thence nothing comes, and it is also in the west." It is perhaps unexpected to find a serious technical work mention the Fortunate Isles so matter-of-factly. This name was probably referring to the Canary Islands, known in Phoenician times but lost track of by Khordâdhbeh's time. It is less surprising to find his book mention Thule, according to Pliny the northernmost land on Earth. Perhaps *Thule* at this point still had its old meaning, "Scandinavia," or its soon-to-be-established meaning, "Iceland." At this time the Ara-

---

[*]Presumably, as postmaster he meant no mail or messages.

bian world was closely in touch with the Scandinavian world via trade routes on Russian rivers. The front map shows how a few important portages across the low continental divide in northern Russia linked the Black Sea to the Baltic and Scandinavia.[4]

880–901: Ottar of Norway. Alfred the Great wrote a historical treatise mentioning Ottar of Norway. Ottar, who said he lived at the northernmost edge of Norwegian habitation, decided to voyage farther northward, to see how far their "island" extended. He and his crew evidently sailed to the northernmost cape of Scandinavia and then eastward as far as the White Sea (see front map). They sailed into it for five days before they were forced to give up and return.

Ottar could not say with certainty if the body of water they had entered was an inlet of the northern ocean or a connection of the sea in the south (Baltic). So the Ptolemaic idea that Scandinavia was an island remained unchallenged. Ottar may even have inadvertently reinforced the idea of an insular Scandinavia. He also tells of a voyage in the opposite direction, through the Skagerrak and Kattegat, past Denmark, and into the Baltic. Ottar did bring back many vivid stories of the Lapps, whom he had met on his northward voyage.[5]

Circa 1000: Greenland-Vinland Voyages. The tenth century ended with the Icelander Erik the Red's discovery and settlement of Greenland about A.D. 984 and, within the next three decades, the discovery and exploration of Vinland by Bjarni Herjolfsson, Leif Erikson and his siblings, and Thorfinn Karlsefni.

The first settlers in Greenland founded the so-called Eastern Settlement (around modern Julianehab), closest to Iceland. (The east coast of Greenland was unapproachable because of pack ice.) Later a Western Settlement (actually to the north) sprang up 400 miles farther from Iceland (around modern Godthab, recently renamed Nuuk). By the thirteenth century the Greenlanders had discovered that sea mammal hunting was better in the north parts of Greenland, at the northern hunting grounds, Nordrsetur ("North settlings"). The location of Vinland has been controversial, and it still cannot be considered to have been identified conclusively.* Nor have any real bound-

---

*In my 1972 book, *Viking America,* I deduced for Vinland a very northerly location, around Ungava Bay, based on the pastureland interpretation of Vinland. The scholarly world has not embraced this theory, but neither has it demolished it. Eugene Fingerhut (*Who First Discovered America?,* Claremont, Calif., 1984, p. 45) first sought to dismiss it but has more recently become open to it (*Explorers of Pre-Columbian America?,* Claremont, 1994, p. 77).

aries of Nordrsetur been established, and we will see that it could have included more than the northwest coast of Greenland.

There is no evidence that the original Norse had any contact with Dorset Eskimos in southern Greenland, which was by then unoccupied except for the Norse (although they did find ruins from earlier Eskimos). During Karlsefni's explorations in Vinland (wherever it was) he encountered Native Americans and carried out trade with them as described in Erik the Red's *Saga*. West of Greenland, Dorset Eskimos coexisted with incoming Thules until the fourteenth century. The descriptions of Vinland as a land of comparatively good qualities competed for attention in Iceland with stories of Greenland's rigors. Those who sought the Isles of the Blessed or the Fortunate Isles were later to seize upon "Vinland the Good" as a kind of earthly paradise.

Little is known about the medieval government in Greenland, although isolated hints suggest it was patterned after that in Iceland. Namely, Iceland developed a highly democratic national parliament (Althing), which governed along with local parliaments and local chieftains, both of which had limited powers. Erik the Red was such a chieftain.[6]

Circa 1000: Anglo-Saxon Map. Although its origin is unknown, the Anglo-Saxon map of the world (Figure 12) has some inscriptions in Anglo-Saxon. It is relatively free of the dulling influence of the medieval wheel map tradition. It nevertheless falls short of Ptolemy's level of accuracy and it needs to be interpreted in the way prescribed at the end of the foregoing chapter.[7]

The Anglo-Saxon map gives a good preview of the confusion that was to come into the cartography of the North with the new discoveries of the Norse. It shows recently discovered Iceland correctly as an island, labeled *Tylon,* to the northwest of Britain and Ireland, and it illustrates the historical transfer of the name Thule from Scandinavia to Iceland (and foreshadows a later transfer to Greenland). However, it identifies the Danish peninsula with Norway ("Neronorroen") and mistreats Scandinavia badly. Admittedly it shows Scandinavia as a "very long and very narrow land," as Ottar had described it, and, incidentally, as an island. But the inscription "Scridefinnas" places the Lapps (then called Finns) at the lower end, closest to the Danish peninsula, where the Norwegians and Swedes should properly have been. Furthermore, the inscription "Island" at the far end of Scandinavia, while to modern English readers meaning a body of land surrounded by water, in Anglo-Saxon meant "Iceland."

One can see through the confusion with a moderate amount of patience.

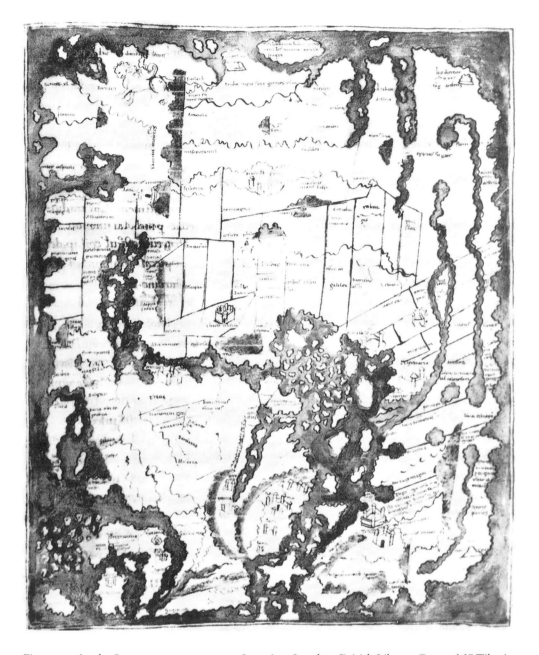

Figure 12. Anglo-Saxon map, ca. A.D. 1000. Location: London, British Library, Cotton MS Tiberius B. V. fol. 56 v. Source: Kamal, *Monumenta Cartographica,* vol. 3, fasc. 1, fol. 545.

The basic actual error was the identification of Norway with the Danish peninsula. Indeed, at that time the terms *Norse* and *Danish* were equivalent, and the British identified all Norse invaders of their islands as Danes. Perhaps the shifting of the name Thule in some way helped cloud the situation. In any case, beyond the identification of Denmark as Norway, everything else follows logically. The Lapps are truly to the north of the Norwegians, and Iceland is truly more remote than Lapland. However, Iceland (Island) is incorrectly placed to the east of Norway, whereas the known sailing directions placed the true Iceland to the west. This cannot be dismissed as mere error, and is similar to the thinking of Adam of Bremen.

Circa 1076: Adam of Bremen. This learned German teacher made no maps, but he wrote a history of the church in the North. That history included verbal geographical descriptions from which Axel Björnbo has synthesized the map shown in Figure 13. While the Icelandic family sagas are assumed to have preserved some history of Greenland and Vinland in oral form from the beginning, Adam of Bremen is the oldest known documentary (hard copy) reference to these countries, recorded about a hundred years after their settlement.

Adam of Bremen was the first to suggest that Scandinavia might be connected to the mainland rather than being an island. However, the suggestion was very indirect; it was implicit in his description, and he gave no references to sources of the idea nor supporting quantitative data.[8]

Note this striking feature of Adam's description (reflected in Björnbo's map): while he mentioned Iceland, Greenland, and Vinland in their correct sequence of remoteness from Scandinavia, he laid out their sequence proceeding north and eastward instead of westward. That the Anglo-Saxon map had the same error suggests that this may have been a systematic and general misunderstanding. One explanation is to postulate an awareness that Greenland and Vinland were not far from continental land. We shall see this emphasized and clarified in later maps.

Adam of Bremen had a penchant for interpreting the meaning of place names, and his interpretations sometimes tended to be whimsical. Concerning Greenland, he said, "The people are bluish-green from the salt water; and from this the region takes its name." Actually Erik the Red, by his own statement, chose the name as a contrast to *Iceland,* to attract colonists to the land. Adam was also the first writer to give an interpretation of the meaning of *Vinland* as "wineland." It was so called, he said, "because vines grow there of themselves and give the noblest wine. And that there is abundance of unsown corn (grain) we have obtained certain knowledge, not by fabulous suppositions, but from trustworthy information of the Danes." Adam's romantic interpretation of the word *Vinland* along the style of the Insulae Fortunatae as an earthly paradise seems to have gained immediate acceptance in Europe. I have maintained elsewhere that the name *Vinland* as applied by Leif

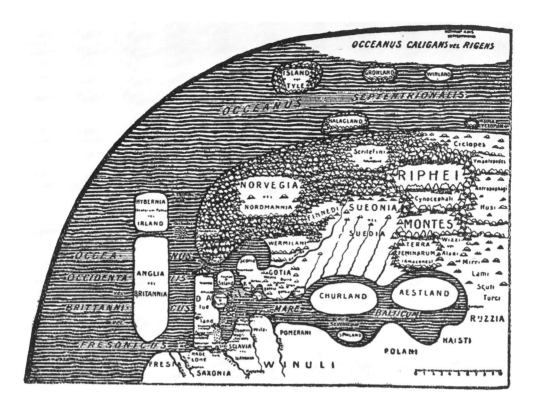

Figure 13. Cartographic reconstruction by Axel Björnbo of Adam of Bremen's geographical conceptions, recorded in 1076. Source: Nansen, *In Northern Mists,* vol. 1, p. 186.

Erikson actually referred to pasture for the Norse cattle and had nothing to do originally with grapes. Folklore is replete with legendary details that have been added by later writers for their own reasons. Indeed, it has even been suggested that those who later put down the sagas on paper could have gotten the wine idea from Adam's writings instead of from oral tradition.[9]

1117–1121: Eric Gnupsson's Search for Vinland. Eric Gnupsson became the first missionary bishop of Greenland in 1112. Presumably he would have traveled about the country organizing priests for a concerted attack against the Aesir faith and the worship of Thor and nature spirits. In either 1117 (according to the Yale Vinland Map) or 1121 (according to the *Icelandic Annals*) he made a trip in search of Vinland. We know no details of this trip, but the Yale Vinland Map's statement that he "remained a long time in both summer and winter" combines with the difference of recorded date to suggest that the trip lasted at least four years.[10]

Gnupsson's voyage is the only recorded voyage to Vinland after the explorations by Erik the Red's family terminated 100 years earlier. The *Icelandic Annals'* use of the word *search* suggests that people no longer frequented Vinland. The trip was evidently remarkable enough for the annalist to give it top billing for that year, although it apparently did not yield any practical results that Greenlanders could integrate into their daily lives. Nevertheless, the geographical understanding of the previous discoveries continued to incite discussion in Europe.

Before 1159: Icelandic Geography. The Icelandic abbot Nicklaus Bergsson probably wrote part of an untitled geographical tract shortly before his death in 1159. It survives in manuscripts from the fourteenth and fifteenth centuries. One passage reads:

> North of Scotland is Denmark. Through Denmark the sea goes into [the countries of the Baltic]. Sweden lies east of Denmark, but Norway on the north. To the north of Norway is Finmark [Lapland]. From thence the land turns toward the northeast, and then it is not far to the east before one comes to Bjarmeland [Karelia, northern Russia]. . . . From Bjarmeland the land stretches to the uninhabited parts of the North, until Greenland begins. To the south of Greenland lies Helluland, next to it is Markland, and then it is not far to Vinland the Good, which some think to be connected with Africa (and if this be so, then the Ocean must come in between Vinland and Markland).[11]

Halldor Hermannsson, having difficulty making sense of this, thought parts of this passage might have been interpolated in the fourteenth century.[12] The most amazing feature of this description is its perfect correctness until Bjarmeland is passed. Then we unexpectedly encounter Greenland as part of the easterly sequence. The correctness about the wastes beyond Bjarmeland shows that people had diligently followed up Ottar's explorations in the North. However, the hypothesis of the one-ocean paradigm leads to a correct interpretation of the latter phrases also. Adam of Bremen had earlier given a convincing description of why the Earth is round, and he had implied that only uneducated people thought otherwise. The author of the tract, whether Bergsson or otherwise, surely held the established concept of a spherical earth and the home circle of the world continent.[13] He envisioned Greenland as extending from that circle at the polar area. Without knowledge of the intervening New World (i.e., its identification with the Old World), he would have presumed any continental land near to Greenland to be arctic Asia. Now, when he goes beyond Greenland his description takes a southerly sequence. That is exactly what it should do on the model of the home circle after passing the North Pole. The circle turns southward after the polar region toward eastern Asia, locating Helluland, Markland, and Vinland there. This could explain why both the Anglo-Saxon map and Adam of Bremen "mistakenly" gave an eastward sequence instead of westward for the Norse discoveries. Whoever conceived this geography seems to have been motivated by information that Greenland and Vinland were near to a continuation of continental land, in reality North America's eastern seaboard.

The suggestion that Vinland is connected with Africa still confronts us. Modern scholars, led by Axel Björnbo, have always interpreted this to mean that the Atlantic is an inland sea, with the outer shore of the Earth's disk being not Europe but some far shore of Vinland.[14] However, Björnbo's conception has no direct support; no ancient authorities have stated such a thing explicitly.* Indeed, the Björnbo picture seems based mostly on the modern academic bias that from Greenland one must go west while going south to Markland and Vinland. In fact there is another way that such a transit could be accomplished conceptually within the disk of the world's lands. That is by taking the global view we used above and seeing Vinland as on the far shore of the great ocean and connected to Africa via eastern and southern Asia. This could be evidence of how far south the Norsemen's awareness extended. Namely, consider South America, which is connected to North America by an isthmus (Panama). Under the one-ocean paradigm the Norse would have had to see South America as Africa, to be reached by rounding eastern Asia, and connected to Africa by an isthmus (Suez).†

The inland Atlantic picture is supported, in the modern mind, by the parenthetical part of the Bergsson quote above, "and if this be so, then the Ocean must come in between Vinland and Markland." However, Nansen has pointed out that the earliest manuscripts do not contain this statement. Perhaps it is a rationalization inserted by some intermediate scholar and copied by later ones, who must not have perceived the description's eastward rather than westward progression along the Earth's disk.[15] Such an eastward mental tour under the one-ocean paradigm may admittedly be a shore-hugging one, but

---

*The Norse term *Hafsbotn,* meaning "bay or gulf of the ocean," was commonly used for the oceanic waters the Norsemen crossed. Some have argued that this is evidence of an inland sea conception of the Atlantic. However, this term applied only to the North Atlantic, which they indeed conceived as a gulf between Greenland and Europe.

†This bypasses a question about whether Ptolemy's Indian Ocean had an eastern connection to the Pacific, but that modern academic question did not exist for the Norsemen, who had never heard of Ptolemy.

it would decidedly involve global thinking. It would also place Vinland "the Good" in the vicinity of the East Asian terrestrial paradise.

Early 1200s: Saxo Grammaticus. The Danish historian Saxo Grammaticus wrote a complete Danish history including geographical descriptions. It clearly illustrates the disparity between continental and Icelandic awareness of Norse activities. Saxo's influence was strong,[*] and a casual statement of his, for which he gave no sources, may have sealed the cartographic destiny of Scandinavia. He described the waters surrounding Scandinavia:

> The [southern] arm of the ocean, which cuts through and past Dania, washes the south coast of Gothia [Sweden] with a bay of fair size [the Baltic]; but the [northern] branch, which goes past the north coast of Gothia and Norwagia, turns towards the east with a considerable widening, and is bounded by a curved coast. This bay of the sea [the White Sea] was called by our ancient primeval inhabitants *Gandvicus.* Between this bay and the southern sea lies a little piece of continent, which looks out upon the seas washing it on both sides. If nature had not set this space as a limit to the two almost united streams, the arms of the sea would have met one another, and made Suetia and Norwagia into an island.[16]

Thus Saxo pronounced Scandinavia not an island but connected to the mainland, although by an isthmus. While the White Sea and the northern arm of the Baltic do close in toward one another, the intervening two hundred miles or more of land hardly "looks out upon the seas washing it on both sides." This description would influence the cartography of Scandinavia's attachment into the sixteenth century.

Early 1200: Heinrich (not) of Mainz. A dedication to one Henricus is associated with this unsigned map (Figure 14), and an eighteenth-century scholar surmised this to refer to a certain churchman from Mainz of ca. 1110. More recent scholars have rejected this and posited a date a century later and a different Henry (perhaps of Huntingdon).[17]

The author followed the formalized wheel map tradition but nevertheless tried within the tradition to represent naturalistic reality. He shows Ganzmir

---

[*]Among other things, he was responsible for relaying the conception of Hamlet from its Icelandic origins to William Shakespeare.

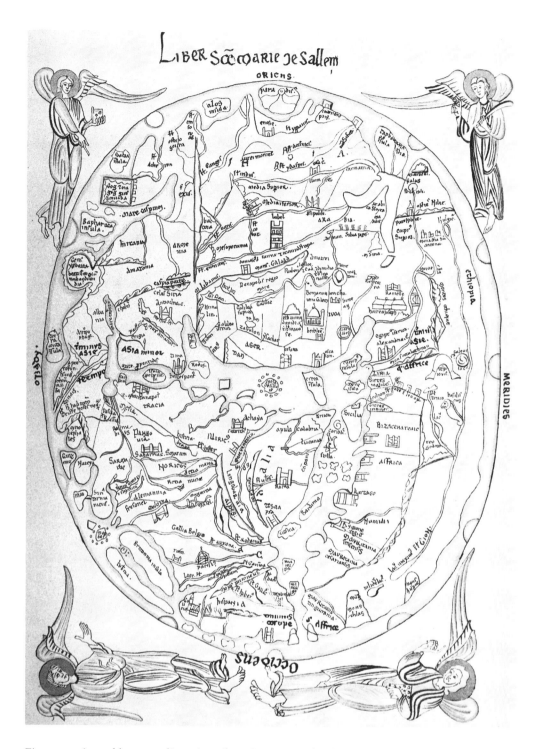

Figure 14. A world map earlier misattributed to Heinrich of Mainz, early 1200s. Location: Cambridge, Corpus Christi College, MS 66, p. 2. Source: Santarem, *Atlas,* pl. 10. Courtesy of Map Division, The New York Public Library, Astor, Lenox and Tilden Foundations.

(Scania, or Scandinavia) and Island (Iceland) as islands connected to the mainland by stylistic isthmuses. This resonates with the thinking of Saxo Grammaticus, but we don't know which came first.

SUMMARY

For more than two hundred years Norse colonists on Greenland went about their business and European scholars went about theirs, with little notice of one another. The few scholars who did come into contact with fringes of information about the Norsemen had an inadequate conception of mainland Scandinavia.

In the Norse world itself, there may have been some applications of the one-ocean paradigm based on early encounters with the American eastern seaboard. However, any such exploration seems to have ended within a few generations after Leif Erikson. At the same time, any contact that the Norsemen had had with native American peoples seems to have been interrupted.

Life in Greenland was very lonely. Individual homesteads were isolated from one another by high mountains and long fiords cluttered with ice. During the major part of its first two centuries of existence, the Greenland settlement was as isolated as it is possible to be while remaining part of the civilized world. Nevertheless, geographical concepts related to their activities filtered into the Western and Central European consciousness.

CHAPTER 7

COMMUNICATION LINKS WITH GREENLAND

THERE WERE TWO FORCES THAT MADE GREENLAND part of the Old World community, in spite of its isolation. These were the Christian church and the royal sport of falconry. Their effect was to make Greenland become known throughout western Eurasia. We will also find that the Greenland colony became an intellectual bridge by making new contacts with the people of America.

1214: Ibn Said. Ibn Said was an Arab geographer who wrote *The Extent of the Earth in Its Length and Breadth.* In describing the islands "around Denmark" (meaning the entire Norse sphere) he provided this image:

> small islands where the falcons are found. To the west lies the Island of White Falcons, its length is about seven days [travel] and its breadth about four days, and from it and from the small northern islands are obtained the white falcons, which are brought from here to the Sultan of Egypt, who pays from his treasury 1000 dinars for them, and if the falcon arrives dead the reward is 500 dinars. And in their country is the white bear, which goes out into the sea and swims and catches fish, and these falcons seize what is left over by it, or what it has let alone. And on this they live, since there are no [other] flying creatures there on account of the frost.* The skin of these bears is soft, and it is brought to the Egyptian lands as a gift.[1]

This Island of White Falcons sounds like Greenland, with the "small northern islands" lying off the coast of the Greenlanders' northern hunting

---

*This surmise of Ibn Said's is incorrect. The Arctic teems with bird life.

grounds. Some writers have suggested that before this time, invading Turks had severed the Scandinavian-Byzantine trade routes across Russia, but a few boats may still have gotten through, and certainly the Mediterranean route was well developed. In any case, the settlements in Greenland had by now grown to a modest prosperity through their specialty trade in falcons, walrus ivory, walrus hide ropes (for ship rigging), and furs.

1240s: **King's Mirror.** The encyclopedic work *King's Mirror*[2] includes a geographical history of the North, principally Iceland and Greenland, in dialogue form. The work was written anonymously shortly before the coronation of King Hakon of Norway in 1247. It clearly describes life in the Norse Greenland settlements at the peak of their flourishing period:

> But in Greenland, as you probably know, everything that comes from other lands is dear there; for the country lies so distant from other lands that men seldom visit it. And everything they require to assist the country, they must buy from elsewhere, both iron (and tar) and likewise everything for building houses. But these things are brought thence in exchange for goods: buckskin and ox-hides, and sealskin and walrus-rope and walrus-ivory. . . .
>
> But since you asked whether there was any raising of crops or not, I believe that country is little assisted thereby. Nevertheless there are men—and they are those who are known as the noblest and richest—who have tried to sow grain as an experiment; but nevertheless the great multitude in that country does not know what bread is, and never even saw bread. . . .
>
> Few are the people in that land,* for little of it is thawed so much as to be habitable. . . . But when you ask what they live on in that country, since they have no grain, then (you must know) that men live on more things than bread alone. Thus it is said that there is good pasture and great and good homesteads in Greenland; for people there have much cattle and sheep, and there is much making of butter and cheese. The people live on this, and also on flesh and all kinds of game, the flesh of reindeer, whale, seal and bear; on this they maintain themselves in that country.

The *King's Mirror* also gives a good account of the difficulties encountered in sailing from Iceland to Greenland:

---

*The means of the widely varying estimates I have read for the peak combined population of the Eastern and Western Settlement range from 5,000 to 10,000, but the median estimate is considerably lower. The most reliable recent estimate is just over 2,000 at the peak.

So soon as the greater part of the sea [Denmark Strait, see front map] has been traversed, there is found a mass of ice as I know not the likes of anywhere else in the world. This ice is some of it as flat as if it had frozen on the sea itself, four or five cubits thick, and extends so far from the coast that men may have four or five days journey across the ice. But this ice lies off the land rather to the northeast or north than to the south, southwest or west; and therefore anyone wishing to make the land should sail round the land [i.e., Cape Farewell] in a southwesterly and westerly direction, until he is past the danger of all this ice, and then sail thence to land. . . .

These ice floes are strange in their nature; sometimes they lie as still as might be expected, separated by cracks or large fiords; but sometimes they move with as great rapidity as a ship with a fair wind, and once they are under way they travel against the wind as often as with it. There are indeed some masses of ice in that sea of another shape, which the Greenlanders call "icebergs." Their appearance is that of a high mountain rising out of the sea, and they do not unite themselves to other masses of ice, but keep apart.

The *King's Mirror* espouses a cosmology of a spherical Earth with the lands, *terra firma,* restricted to a disk-like cap or "home circle." The author thinks that Greenland is "mainland, and connected with other mainland." Referring to one authority, he says that Greenland "lies on the extreme side of the world to the north," and that he "does not think there is land outside the disk of the world's land beyond Greenland, only the great ocean which runs round the world; and it is said by men who are wise that the strait through which the empty ocean flows comes" into the disk or home circle "by Greenland, and into the gap between the lands, and thereafter with fiords and gulfs it divides all the countries." To make sense of this, I quote from an Icelandic manuscript called *Gripla,* whose date of origin is unknown but which was copied in the early seventeenth century. It takes up from where the *Icelandic Geography* of 1159 (whether by Bergsson or later interpolation) left off: "thence it is a short distance to Vinland the Good, which some people think goes out from Africa. Between Vinland and Greenland is Ginnungagap; it proceeds from the sea that is called Mare Oceanum, which surrounds the whole world."

To make sense of this last passage, we must make a further digression, into Norse mythology. The Icelanders had been Christian for centuries, but in the thirteenth century their rising interest in their history also encompassed the ancient religion. They gathered on paper all the old myths they could and did their best to understand the meanings. The word *Ginnungagap* is involved in the pagan creation story. As with most creation stories, there was a preexisting structure before the peopled world was created. There existed a place called Nifelheim and, far away, another place, called Muspelheim. The great void between them, which was later to be occupied by the world of men, was called Ginnungagap. Once created, the peopled world became Midgard. As thirteenth-century scholars digested these stories, they engaged in the familiar practice of attributing myths to primitive interpretations of real world phenomena. They put mythical places like Giantland on the real map, not necessarily because they believed it but because they thought the ancients believed it. They interpreted Nifelheim, with its described frigid conditions, as the Arctic and they interpreted Muspelheim, with its described burning cinders, as the tropics.

Ginnungagap had an aspect that led to multiple interpretations of it. This gap was not perfectly filled by the creation of Midgard. It was possible to conceive of an edge or holes where Midgard met Nifelheim, and there the ocean might be expected to be running off into remaining bits of Ginnungagap. Thus, thirteenth-century and later maps and scholars began to label various known whirlpools as "Ginnungagap."[3] Such an interpretation was made easier by an etymological ambiguity of the Old Norse word *Gunnungagap*. Jan de Vries has shown that the likely contextual meaning in pagan days, when the belief was current, was "magically empowered empty space," reflecting a sacred portent of the world to come. However, later scholars almost surely seized upon a grosser surface etymology that permitted the interpretation "gaping mouth" or "yawning chasm." This was perfect for generating the whirlpool interpretations.[4]

Now we can return to *Gripla* and its statement that Ginnungagap lies between Greenland and Vinland. There is of course no whirlpool between Greenland and temperate North America (although one later Icelandic map by Jón Guðmundsson[5] shows a whirlpool in what may well be Davis Strait, apparently motivated by the *Gripla*). However, the indrawing channels of the Arctic Archipelago and certainly of Hudson Strait can be construed as lying between Greenland and Vinland. The currents of Hudson Strait would have been known ever since Leif Erikson's time and could easily have deserved the reference to Ginnungagap. And Hudson Strait indeed does provide an entrance to the presumed world disk that goes, as the *King's Mirror* says, "into the gap between the lands, and thereafter with fiords and gulfs it divides all the countries" (see front map).

This interpretation seems superior to Björnbo's inference of an inland Atlantic, which has no other support.

Throughout 1200s: Saga Writing. Since their earliest days, occasional Icelanders with trading or religious contacts must have witnessed writing in Latin on parchment. By the 1100s a movement was developing to promote writing on parchment in the Norse language. This movement became so popular that by the thirteenth century Iceland had established a basis for an extraordinary national literature. This literature naturally turned to the verbal traditions for inspiration, and the sagas as we know them were first recorded at this time.

The sagas are classified according the period during which their stories take place, rather than the period during which they were written down. The classes are (1) the *Fornaldarsögur* or "old times sagas" (also called legendary sagas), which depict heroes and events outside of Iceland and prior to the settlement of Iceland; (2) Icelanders' sagas, or family sagas, which tell the stories of early Icelandic families and activities constituting the classical Saga Age up until the time writing became popular; and (3) contemporary sagas, which describe events contemporary with the era of writing. Most of the family sagas, including the so-called Vinland Sagas, are believed to have been first written down during the thirteenth century. The currently existing manuscripts are copies many times removed from those now-lost originals.

Controversy has always filled the study of the sagas. Here the most interesting question is that of historicity. Were the written sagas truly based on oral tradition rather than invented literature, and to what extent did oral tradition itself preserve the truth until recorded on parchment? The fashion of belief has swung from gospel acceptance of the written sagas, which reigned around 1800, to their complete rejection as inventions, around 1900. By the middle of the twentieth century, scholars had become thoroughly perplexed by the lack of any conclusive evidence in either direction coupled with a plethora of inconclusive evidence in both directions. They henceforth refused to discuss at all the matter of historicity, concentrating only on literary aspects.[6] This situation cannot remain permanently, but it will probably not change until some totally new kind of evidence or a new interpretation of existing evidence is found. One dramatic example of this occurred at the end of the twentieth century and concerned a saga describing one Egil Skalla-Grimsson's physique, formerly thought to be quite fantastic. Jesse Byock has now shown medical reasons to believe that the description was accurate in every detail.[7] Meanwhile, Jônas Kristjansson has argued for some return to historicity.[8]

While Iceland developed a glorious literary culture, we have no awareness of any independent literary development in Greenland. All the knowledge we have about activities in Greenland was recorded in Iceland. Nevertheless, several writers have deduced from the different approaches to the same subject matter in the Saga of Erik the Red and the Tale of the Greenlanders that there were two oral traditions, one preserved in Iceland the other in Greenland. There is also no knowledge of whether or not acquaintance with the Vinland Sagas ever spread to other countries or if their substance was ever translated into Latin.

1260?–1264?: **Historia Norwegiae.** A general history of Norwegian activities, *Historia Norwegiae* was written by an anonymous Norwegian author. Scholars differ on its date, some placing it as early as 1180–90 instead of 1260–64. In it appears the oldest surviving explicit reference to the Greenlanders' northern hunting grounds, Nordrsetur, the region starting to the north of modern Godthaab (Nuuk). The same passage includes the first new reference to contact with the Eskimos after a long historical silence on the subject of natives following Karlsefni's encounter with Skraelings in Vinland 240 years before 1260: "On the other [i.e., west] side of Greenland toward the north there have been found by hunters certain small people whom they call *Skraelings;* . . . they have a complete lack of the metal iron; they use the tusks of marine animals for missiles and sharp stones for knives."[9] That this was truly a rediscovery of the Skraelings is evidenced by the intense interest in them shown by resident Greenlanders on the later voyages of 1266–67, described in the discussion of the letter from Haldor to Arnold, immediately below. This encounter in the middle of the thirteenth century is consistent with the known history of the demise of the Dorset Eskimos and the rise of the Thule Eskimos. The exact chronology of the Thule migration is uncertain[10] but the culmination is well established. After leaving their original home in northern Alaska, in an early phase of migration, the Thule crossed Canada and the Arctic Archipelago. In a perhaps later phase, having the advantage of an already settled region to pass through, they migrated all the way from western Alaska to Greenland, entering via Ellesmere Island (see front map). They then migrated southward, along both coasts of Greenland, reaching the halfway point of Greenland's west coast around the middle of the thirteenth century.

Greenlandic history's long silence on the Skraelings begs to be accounted for. This period of silence about the Skraelings coincides with the period of silence about commerce with Vinland. Once contact with the Thules was established, however, the process accelerated, and archaeologist Robert

McGhee is tempted to speculate that the Norsemen became a magnet for the late-phase Thule migration.[11]

1266–1267: Letter from Haldor to Arnold. There exists an account of two exploratory voyages, one by laymen in 1266 and a follow-up by priests presumably the next year. The account was summarized by Bjørn Jónsson from a now-lost letter from the Greenlandic priest Haldor to his colleague priest Arnold, shipwrecked and stranded in Iceland. Jónsson's summary reads in part:

> That summer [1266] when Arnold the priest went from Greenland, and [his party] were stranded in Iceland at Hitarnes, pieces of wood were found out at sea [near west Greenland, not Iceland] which had been cut with hatchets and adzes, and among them one in which wedges of tusk and bone were embedded. The same summer men came from Nordrsetur, who had gone farther north than had been heard of before. They saw no dwelling-places of Skraelings except in Kroks-fjordarheidr and therefore it is thought that they [the Skraelings] must there have the shortest way to travel, wherever they come from. After this the priests sent a ship northward to find out what the country was like to the north of the farthest point they had previously reached; and they sailed out from Kroksfjordarheidr, until the land sank below the horizon. After this they met with a southerly gale and thick weather, and they had to stand off. But when the storm passed over and it cleared, they saw many islands and all kinds of game, both seals and whales, and a great number of bears. They came right into the ocean's ultimate parts and all the land then sank below the horizon, the land on the south and the glaciers; then there was ice to the south of them as far as they could see; they found there some ancient dwelling-places of Skraelings, but they could not land on account of the bears. Then they went back for three days, and they found there some dwelling-places of Skraelings when they landed on some islands south of Snaefell. Then they went south to Kroksfjordarheidr one long day's rowing, on St. James Day; it was then freezing there at night, but the sun shone both night and day, and, when it was in the south was only so high that if a man lay athwartships in a six-oared boat, the shadow of the gunwale nearest the sun fell upon his face; but at midnight it was as high as it is at home in the Settlement when it is in the northwest. Then they returned home to Gardar [the bishop's seat in the Eastern Settlement].

This tempting collection of astronomical observations on the follow-up voyage should allow one to calculate the latitude of the unknown Kroksfjordarheidr, and it was surely stated with such use in mind. The priests of Greenland who sent this ship were carrying out a methodical exploration. But the height of the gunwale above the deck of the particular kind of six-oared boat they were using is no longer known.[12] Disregarding a variety of speculative attempts to fill in these unknowable parameters, one useful fact is certain. Some part of the sun's disk was visible above the horizon at midnight on Saint James Day. The Feast of Saint James is on July 25th, but by 1267 the Julian calendar had slipped eight days out of synchronism with the seasons as we know them. For astronomical purposes Saint James Day in 1267 corresponds to our Gregorian August 2nd. On this day the center of the zenith sun is $18°$ north of the equator and would be visible across the pole as midnight sun at anywhere above $72°$ north latitude. The solar disk is about one-half degree in diameter, so, allowing for some refraction, the top limb of the disk would be visible at about $71\frac{1}{2}°$ and north. Thus Kroksfjordarheidr must have been at least that far north.

One can further delimit this latitude range by the statement, "but at midnight it was as high as it is at home in the Settlement when it is in the northwest." At home in Gardar, at $61°$ N, the only time the sun reaches the northwest at all is around the longest day of summer, Gregorian June 21st. When it then reaches the northwest, between 8:30 and 9:00 P.M., it has already set behind a high rock ridge running along the west of the plain of Gardar, on which the cathedral buildings were situated (Figure 15a). Thus, from the cathedral ground itself one can *never* see the sun quite in the northwest. However, a little to the north of the cathedral the rock wall is broken by a pass the Norsemen called Eid (Figure 15b). They used this pass to travel overland to the Eiricksfjord shore and thence across to Brattahlid, Erik's home. This pass lies exactly northwest from the cathedral anchorage at Gardar, certainly a spot of practical importance second only to the cathedral church itself. When the summer sun sets in northern latitudes, it does not set perpendicularly to the horizon but at a grazing angle headed northward. After a few minutes the sun reemerges in the pass from behind the hill. From the anchorage it can be seen exactly in the northwest (Figure 15c) before it sets again.

I personally found this visual experience quite memorable. The Gardar priest could easily have called it to his colleagues' minds simply by the phrase "when the sun is in the northwest at Gardar." At this station on June 20th the maximum elevation of the sun above the actual horizon was about three solar diameters, or $1\frac{1}{2}$ degrees. This introduces some ambiguity into the phrase, but it does set a limit.

By this interpretation, the priests in 1267 could not have been more than 1½ degrees of latitude beyond their 71½° minimum limit. Within these limits is a major peninsula of the west coast, Svartenhuk, the last major glacier-free area as one goes north (Figure 16). There is no evidence that the glacier's configuration was significantly different then than now. Beyond Svartenhuk the glacier extends directly into the sea and there are no more fiords at all to satisfy the one mentioned in the description. The Old Norse name *Kroksfjord* means "crooked" or "hooked" fiord. The inner fiord of the Svartenhuk peninsula, modern Uvkusigssat, fits that name perfectly.* After a remarkably long and wavy inlet (more than 40 miles) this fiord suddenly makes a 90° turn to reach its moraine- and stream-filled head against the glacier 8 miles still farther. The name *Kroksfjordarheidr* means the "heaths of Kroksfjord" and evidently refers to some extensive tract of flat wasteland in the vicinity of Kroksfjord. On the west coast of Svartenhuk, at 71°35′ N, there is a broad flat plain, in Eskimo called Narssauq, which extends ten miles back from the coast. There are numerous less spectacular heaths, but all are exceptionally flat and extensive, given the otherwise mountainous Greenland. Just north of the large heath is a fine harbor (Svartenhavn), which even in the twentieth century was the focus of several sledge routes and contained a travelers' house. If this is truly the vicinity of Kroksfjordarheidr, then it provides a reference point from which to trace backwards the return voyage described in the priest's letter.

The "islands south of Snaefell" were one long day's rowing north of Kroksfjordarheidr, which Nansen has suggested to be about 40 miles. At just this distance north of Svartenhavn begins an expanse of over a hundred small islands, unique to the Greenland coast, which extends up the coast to nearly 75° north. The name *Snaefell* (snow-peak) was used by Norsemen from Iceland to the Isle of Man. It was always used for an isolated and widely visible peak that was especially prominent in its surroundings. In this neighborhood of Greenland the most spectacular candidate for such a name would be the peak of the island Qaersorssuaq, at 72°42′ N, which rises straight out of the sea to over 3,400 feet within a mile. If there is any place on this part of the Greenland coast where the name *Snaefell* would be properly used, it would be here. Somewhere between here and the southern end of the island chain, the priests' crew evidently began their long day's rowing, after finding some recently occupied dwelling places of Skraelings to report.

---

*Kroksfjord was previously often guessed to have been Vaigat, above Disco Island, but the identification here fits much better with other parts of the story.

a

b

Figure 15. (*a*) View in modern village at Gardar, from the site of the cathedral ruin, looking directly west, towards the ridge behind which the sun has set at solstice. (*b*) View from same spot, looking northwest, towards Eid Pass. (*c*) View northwest into Eid from a closer location at anchorage, after the sun has moved along the horizon.

c

They had reached those islands after three days of sailing from somewhere, but the location of that somewhere is hardly obvious from the text. To try to understand it, we may go back to the beginning of the letter. There one sees in the second and third sentences that the laymen's first voyage (1266) had searched all around Nordrsetur. They had gone farther north than anybody had gone before, to find traces of the origin of the Skraelings whose dwelling places had been seen on Kroksfjordarheidr. The situation was curious indeed to the Norsemen. They knew that the Skraelings could not have come from the south, which they themselves occupied, and they could find no traces of them in the north, where the glacier covered the seacoast. They further knew that the Skraelings could not have come down from the inland ice in the east, for nothing could live there. Only one possibility seemed to remain—the west. That possibility was strongly suggested, because, during the same summer, pieces of tooled wood with embedded artifacts had been encountered out at sea (Baffin Bay/Davis Strait). Presumably these Skraelings had come from some unknown place out west across the water. Then the Norsemen would have concluded that the shortest way to travel from that place to Greenland must have been such as to have them arrive in Kroksfjordarheidr. Thus, to find this place, one should sail west out from Kroksfjordarheidr.

The expedition the priests (Haldor's Greenland colleagues) organized the following year seems to have done just this. They sailed out into Baffin Bay

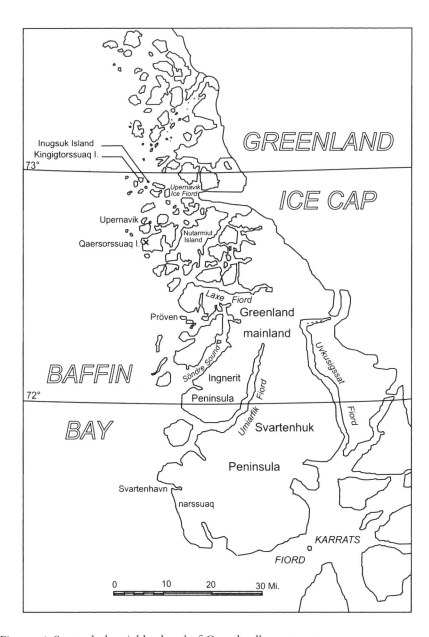

Figure 16. Svartenhuk neighborhood of Greenland's west coast

but perhaps kept a northerly component with hopes of also exploring north of the previous point. After sailing out of sight of land, however, they encountered a southerly storm. Exercising whatever control they could to regain land, they ended up somewhere among the game-teeming islands that dot the Greenland ice-coast all the way up into Melville Bay. They were now perhaps even further north than the previous explorers had gone and still saw no traces of Skraelings. So they sailed straight out into the water, beyond sight of land and the ice cap. At this point they encountered ice to the south of them as far as they could see. This would have been the Baffin Bay "middle ice" pack (see front map). After this experience at sea they evidently came to some part of Baffin Island or the Arctic Archipelago. This was no mere island out in Ginnungagap or Hafetsbotten, but, to the Norsemen, the original home of the Skraelings, a land containing traces of ancient dwelling places. It is difficult to imagine their frustration, after so much investment, at not being able to land "on account of the bears." Polar bears always deserve caution, whether ashore or afloat, but the Norsemen did carry crossbows. The bears may have given the priests a convenient means of saving face, being just as fearful of encounters with these strange new people.[*]

Since the sun shone here twenty-four hours a day, the only reasonable way to proceed homeward was to sail twenty-four hours a day while favorable winds lasted. At the average headway of 5 knots made by Norse ships, the reported three days would be exactly correct for a voyage from, say, Bylot Island to the islands around Upernavik.

The most important historical aspect of this letter is not just the specific identification of the exact area the priests visited. At least as important is the intense interest they showed in the new contact with the Skraelings. Their interest was not likely anthropological diletantism. The priests' business was to assure devotion to Christ in a people who still had a memory of Odin and his pantheon. They had to know everything they could about a heathen people who demonstrated a clear superiority in dealing with Arctic life. The letter makes it plain that the priests organized the expedition for the express purpose of finding out where the Skraelings came from and what their distribution was.

The letter also demonstrates that writing on parchment was practiced in Greenland in 1267. One would normally expect that such an incidental communication would have either been memorized and relayed orally, then forgotten, or written in runes upon a temporary wax tablet. However, that it was preserved long enough to become incorporated into later documents suggests a more permanent record. According to Bjørn Jónsson, who transcribed the existing record in the seventeenth century, "This account was written from Greenland by Haldor the priest to the Greenland priest Arnold, who had become chaplain to King Magnus Hakonsson aboard the mer-

---

[*]Once contact was fully established, the Norse found that the Eskimos posed no physical threat to them. There is no clear record of any on-going Norse-Eskimo strife. The Norse looked upon the Eskimos as merely primitive; the very word *Skraeling* has connotations of "wretched."

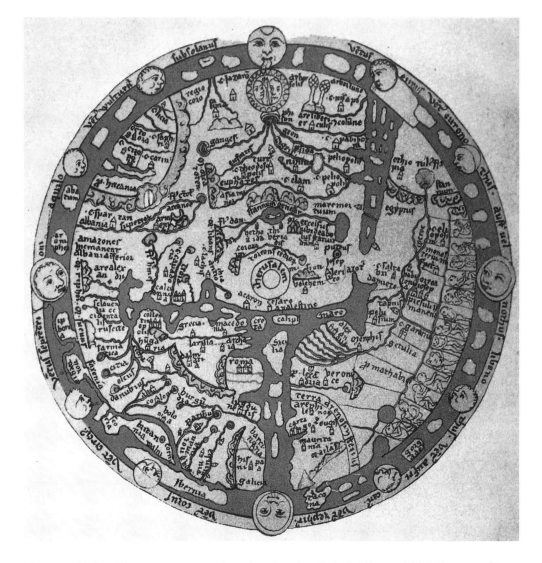

Figure 17. Psalter Map, ca. 1250–1300. Location: London, British Library, Add. MS 28681, fol. 9r. Source: Miller, *Mappaemundi,* Heft 3, Taf. 3. Courtesy of Map Division, The New York Public Library, Astor, Lenox and Tilden Foundations.

chantman that brought Bishop Olaf to Greenland." Jonsson specifically used the verb *rita* without the modifier *snída,* which in Old Norse explicitly differentiated writing on parchment from runic or verbal communication. The exciting written information may thus have reached further than Iceland. At about this same time Roger Bacon, in his *Opus Magus,* resurrected Pliny's statement about people called Hyperboreans and Arimphians who lived right up to the pole.

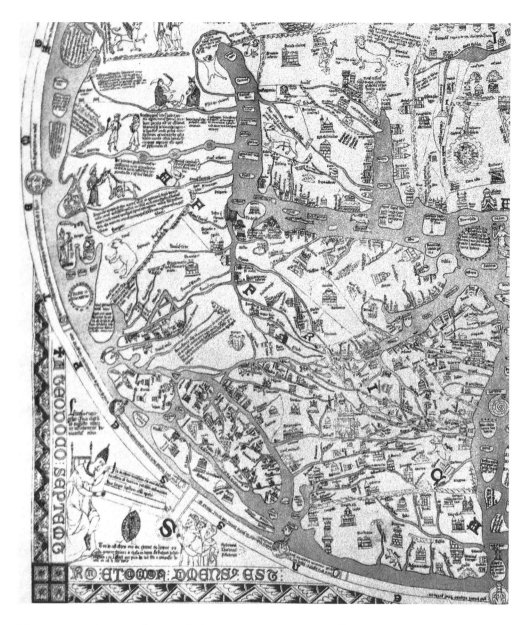

Figure 18. Northwest in the Hereford map, 1280. Location: Hereford Cathedral. Source: (From lithography by Konrad Miller), Kamal, *Monumenta Cartographica,* vol. 4, fasc. 1, fol. 1077.

Circa 1250–1300: Psalter Map. This map (Figure 17) is in the same tradition as that of Heinrich (not) of Mainz, but it shows evidence that the Norsemen's recent contact with the migrating Eskimos caused geographical puzzlement in Europe.[13] While the Psalter Map still shows the two lands that Heinrich's connected by isthmuses to the mainland, it has completely dropped the identity of Iceland. It has moved Norwegia to the former position of Iceland and filled the gap by turning to ancient mythology; the upper peninsula is labeled Hyperboria. It has added a new island beyond the North Pole and revived the ancient myth of the Arimphians. This new island could be the first cartographic trace of new Eskimo information on lands beyond Greenland.

Circa 1280: Hereford Map. Drawn by Richard de Haldingham, the Hereford Map (Figure 18) is of the same school as and approximately contemporary with the Psalter Map. It is covered with depictions of a variety of medieval superstitions, and its formalism is much less strict than that of the Psalter Map or Heinrich's map. Scandinavia is drawn quite naturalistically, with its two southern bulges and a connection to the mainland. Just below the Scandinavian peninsula are two islands labeled (nearly illegible at this scale) Ysland (Iceland) and Ultima Tile. When an island had two names, medieval cartographers sometimes duplicated it under each name. However, some cartographers also used the name Ultima Thule for Greenland, which was even more ultimate than Iceland. If such is the case here, it would illustrate how a very large land far from Scandinavia could be misunderstood in the rest of Europe as part of Scandinavia.[14]

SUMMARY

Greenland, the most provincial outpost of all Europe, after three centuries of inbreeding and monotony, suddenly underwent the most electric of all experiences—contact with a new people and new culture. The Greenlanders' first reaction to the migrating Eskimos was intense curiosity, and there is no evidence of any overt hostility between the two peoples at this time.

News of these curious beings apparently reached Europe and rekindled interest in ancient myths of northern Eurasian peoples. Meanwhile, in Greenland itself a possibility arose for the Norsemen to learn many new things from the Eskimos, including mapmaking.

CHAPTER 8

THE UNSEEN BRIDGE

THE GREENLANDERS CONTACT WITH THE MIGRATING ESKIMOS literally opened a whole new world to them. Many traces of the Norse discoveries found their way into European documents. But, ironically, as the Greenlanders turned their attention westward, the southern European world was turning its eyes to other directions and found new reasons to ignore the Greenlanders. I hypothesize that the result was an information transfer, one that is poorly documented but, as we shall see, foreshadowed some of the startling misunderstandings discussed earlier.

1285–1295: Nyaland and Duneyar. The *Icelandic Annals* for the year 1285 contain an entry that in that year two priests, Adalbrand and Thorvald, discovered the "New Land" (Nyaland) and the "Down Islands" (Duneyar). The later *Landnamabok* contains the only hint to the location of these discoveries: "Well-informed men have said that southwest is the course from Krysuvikur Bergi [near the southwesternmost point of Iceland] to Nyaland." People must have long before discovered all the islands in the vicinity of Iceland, in the four hundred years since Iceland's original settlement. The next nearest land in a southwestern direction from Iceland is the uninhabitable east coast of Greenland. However, in 1289 King Eric of Norway encouraged his subject Rolf to organize colonizers for Nyaland, making East Greenland an unlikely candidate. Possibly something beyond Greenland? The sailing directions for anywhere related to the Greenland settlements start out with "southwest."

Most likely the small clerical community, including the priests Adalbrand and Thorvald, would have heard about their Greenlander colleague Arnold who had been shipwrecked and stranded in Iceland in 1266 (page 103). A survivor of a shipwreck is always a big news item. An even bigger news item would have been the letter about Skraeling contacts received by Arnold from Haldor. No record exists of Arnold's activities after receiving the letter, but the natural reaction would have been to return and join the explorations. The natural reaction of many others in Iceland must have been to want to join. Adalbrand and Thorvald may have been leaders in such a movement. Rolf pursued recruiting participants vigorously until his death in 1295, by which time he had acquired the nickname "Land-Rolf."[1]

It is possible and even reasonable to interpret the course description in the *Landnamabok* as simply saying that Nyaland was somewhere in the new lands beyond Greenland; perhaps it was even used as a generic term. Some writers have tried to extend the course line directly to Newfoundland. They attribute Cabot's later choice of that name to foreknowledge of Norse explorations and a transliteration of *Nyaland*. While there may be some later evidence in such a direction, the immediate chronology speaks against it. Activities at this time seem to have been north and west of Greenland, not south. Indeed, the name Down Islands would have been quite appropriate for the islands extending up toward Melville Bay, which teem with bird life.

Before 1300 (?): **Geographia Universalis.** This geographical encyclopedia was probably written in England during the thirteenth century. At that time and place there was much interchangeability about Germanic and Romance *V*'s, *F*'s, and *W*'s, and there has been uncertainty as to whether the following entry refers to Vinland or Finland.

> Wynlandia is a country along the mountains of Norway on the east, extending on the shore of the ocean; it is not very fertile except in grass and forest; the people are barbarously savage and ugly, and practice magical arts, therefore they offer for sale and sell wind to those who sail along their coasts, or who are becalmed among them. They make balls of thread and tie various knots on them, and tell them to untie three or more knots of the ball, according to the strength of the wind that is desired. By making magic with these [the knots] through their heathen practices, they set the demons in motion, and raise a greater or less wind, according as they loosen more or fewer knots in the thread, and sometimes they bring about such a wind that the unfortunate ones who place reliance on such things perish by a righteous judgment.[2]

The statement that this country lies "along the mountains of Norway on the east" has caused some writers to accept it unquestioningly as referring to Finland. Perhaps partly because of the similarity of names, this is exactly where many medieval geographers had also placed Vinland. This placement occurred as early as Adam of Bremen in 1076 and was practiced by other northern geographical writers. We have seen that it is a natural and proper placement of Vinland under the one-ocean paradigm.

In fact, the Norse rediscovery of the Eskimos and their lands would likely have led to a renewal of interest in Vinland information toward the close of the thirteenth century. The description of the magic thread and knots sounds a little like cat's cradle games and is suggestive of a typically Eskimo art form, making string figures. There was a superstition connecting string figures with a magical quality affecting the wind. It was taboo to make string figures under the light of the sun. If this taboo were violated, bad, stormy, or especially windy weather and evil spirits would be released. Then only a shaman (Eskimo medicine man) could undo the accidental release of these forces. It is not difficult to imagine both sides capitalizing upon such a belief as that described in the quotation, in any close association between Norse sailors and the Eskimos.

1294: Greenland's Trade Declines. From its founding in 984 by Erik the Red, the Greenland settlement had been, like Iceland, a free state with no political ties to Norway. However, during the impressive reign of the Norwegian king Hakon Hakonsson, the Greenland settlements became persuaded that they would gain many advantages under the Norwegian crown. Sometime around the 1250s, Greenland more or less voluntarily succumbed to pressures to become a Norwegian province. In Iceland political power had gradually shifted from the assemblies into wealthy chieftain families, who then became engaged in complicated intrigues. Iceland's merging into Norway shortly followed. Hakon's successors, following the world trend, became involved with the political plum of colonial trade. In 1294 a new law essentially allowed the king to specify which merchants could practice colonial trade, to the exclusion of others. The "in" merchants naturally found it to their advantage to concentrate on the nearer colonies (Hebrides, Orkneys, Shetlands, Faeroes, Isle of Man, and shortly, Iceland). Soon the only communication between Greenland and Europe was the annual visit of the royal knarr (merchant ship) to supply absolute necessities (see page 98). Icelanders, and certainly Greenlanders, had meanwhile exhausted any timber stocks to build their own new ships.[3]

In some years the knarr was wrecked and never reached its destination. Then the Greenlanders had to learn how to adapt to life without it and its cargo. This adaptability served them well over the next century. The intervals between visits by the knarr became a decade or more, as pirates swept the sea and the Hanseatic League began dominating Norwegian trade.

We can only speculate about the means of the adaptation. One can expect that the situation would have led to some kind of relations with free-traders, privateers, and perhaps pirates. An even more valuable mode of adaptation would have been to learn whatever they could from the Eskimos, regardless of the church's attitude toward contact with heathens. Nothing of this sort is mentioned at this time in official clerical or political communications, but official mention should not be expected, since contact with either of these groups would have involved an illicit element. Later evidence will suggest that such relations did occur.

Circa 1298: Marco Polo. While the Norsemen were establishing contact with a new people to the west, two Venetian merchants were, by accident, establishing such contacts to the east. The father, Niccolo, and uncle, Maffeo, of the child Marco Polo had an established trading business in Istanbul, and it led them, in the early 1260s, to the Caspian Sea region. There, in unfamiliar territory, they suddenly found their return route blocked by the outbreak of local wars. After more traveling about, during which they learned the Tartar language, they befriended another group of travelers from a distant land. These new friends convinced the Polos that their command of the eastern language was a valuable combination with their western backgrounds. The travelers invited the Polos to visit their own ruler far to the east and investigate further trade possibilities. Reaching this land "far to the east" turned out to require a year of constant traveling. Their new host turned out to be the Great Kublai Khan of Cathay.

It was soon recognized that direct trade was impractical, but the Khan had another interest for which he cultivated the Polos' friendship. He had heard of Christianity and the Pope at Rome and desired to have missionaries sent to Cathay. For the benefit of their souls, as well as, presumably, their pockets, the Polos were willing to attempt this gigantic round trip. They made their way back to Venice in three years. For reasons beyond their control, they were unable to return to Cathay with the desired missionaries, but by 1271 they had decided to return to the Khan with another request of his—some oil from the lamp of the Holy Sepulcher. They also took with them the then seventeen-year-old Marco. The journey itself took over three years, and once there, they remained for seventeen years. Marco traveled all about the Orient on various business for the Khan. Their position of security was threatened, however, as their only, if powerful, friend, Kublai Khan, approached old age. They managed a gracious exit shortly before his death, returning to Venice in 1295 after a sea voyage over the China Sea and Indian Ocean.

Back home they found themselves in the middle of a local war, during which Marco was captured, then imprisoned until 1299. During his imprisonment he dictated his famous book describing in great detail all the fabulous sights he had seen in the Orient.* The romance of the story appealed to general readers, but the geographical details, to be examined below, did not have their impact until much later. Meanwhile, other people continued the contact with the East until about 1340, when the Tartar empire fell and a new, hostile government ejected all foreigners.[4]

Early 1300s: **Rymbegla** *Tract.* A discussion by an anonymous Icelandic author about the calendar and the seasons also contains occasional geographical digressions. One such section (chap. 7, par. 50) reads as follows:

> There are those who assert that the ocean stretching northward of us up to the pole, where the branches of the outer ocean run together, is full of perpetual solid ice. The truth of this is hinted at by a verse of Master Johannis of Pavilla in his book, entitled *Architenius.* It goes:†
>
>> Tre forma terris fecis supernatat unde
>> Curva superfacias terramque complectitur arcu
>> Superfecta maris prohibente litore supera.
>
> Which is nearly this in translation: "The shape of the Earth is round, and it carries over itself the rounded surface of the waters and the sea; the ring of the sea embraces the world continent in its arc, which some shores interrupt."
>
> There are those who think that this should be interpreted that he says that land lies under the polar star, and that the shores there prevent the ring of the ocean from coming together. This is in accord with those wise olden times sagas which show that one can go, or that men have gone, on foot from Greenland to Norway.[5]

---

*There has been doubt and controversy about whether Marco Polo ever actually made the trip he describes but perhaps based his descriptions on other sources. This question is irrelevant to the history we are studying, which is concerned with the undeniable effects of his book.

†This Latin verse is isolated in a text which is otherwise strictly in Old Norse, and its corrupted form is probably attributable to scribes rather than to Johannis de Pavilla. Scholars have reconstructed the probable original thus:

> Terrae forma teres, facies supernatat undae;
> Curva superficies terram complectitur arcu
> Superjecta maris, prohibenti litore supra.

The reasoning involved in interpreting this verse relates to the one-ocean conception of the globe (see page 18). The *Rymbegla* author suggests a departure from circularity of the world continent with Greenland projecting out over the pole from a root in arctic Asia, just as Bergsson envisioned.

The book *Architenius* is no longer known,[*] but the sagas referred to probably include that of the Greenlander Halli Geit. This saga itself is no longer known in its original form but was summarized by a seventeenth-century scholar thus: "He alone succeeded in coming by land on foot over mountains and glaciers and all he wastes, and past all the gulfs of the sea to *Gandvik* [the White Sea] and then to Norway. He led with him a goat, and lived on its milk; he often found valleys and narrow openings between the glaciers, so that the goat could feed either on the grass or in the woods." It is out of the question that anybody could have actually done this in the real world. However, what seem like tall tales are sometimes based on a grain of truth, with embellishments making them tall. A real Halli living off seals, not a goat, along the Canadian arctic coast could have engendered such a story under the one-ocean paradigm. The suggestion of a *westward* course in first mentioning Gandvikus then Norway ties in with the idea that Greenland was adjacent to eastern Asia. A real westward journey from Greenland, retracing the migration route of the Eskimos (with Eskimo guidance), is not hard to contemplate if one is familiar with Eskimo life. If the journey went all the way west to the homeland in Alaska, that would have corresponded conceptually to Scandinavia. It was at the northwest corner of a continent, and there was supposed to be only one continent. Halli and the Greenlanders may have themselves believed that the Eskimos were Laplanders (or Karelians) who had found their way overland from northern Scandinavia (i.e., Alaska).

Circa 1300–1330: Carignano's Mariners' Chart. Carignano's was one of the earliest mariners' charts to have survived into modern times. Then a bombing raid in World War II ended its existence. It was drawn by a Genoese priest shortly after the technique of such so-called portolan charts became standardized. People were now to use such works (if perhaps not this particular one) for navigation rather than mere illustration.[6]

Carignano's chart shows some amazing examples of accuracy, not the least

---

[*]A marginal gloss in a contemporary manuscript on a similar subject refers to "the philosopher Architenius." See Lynn Thorndike, *The Sphere of Sacrobosco and Its Commentators,* (Chicago, 1949), p. 436.

Figure 19. Mariners' chart by Giovanni Carignano, ca. 1300. Source: Theobald Fischer, *Raccolta*, pl. 3. Courtesy of Map Division, The New York Public Library, Astor, Lenox and Tilden Foundations.

of which is his rendition of the Danish peninsula. His wind-rose framework ends just north of Denmark, but beyond the framework he appears to have depicted Scandinavia with a level of accuracy (see front map) that was unknown previously and indeed would not recur for centuries to come. What appears to be the Baltic runs off to the right at about the correct angle; the Skagerrak and Kattegat appear to be present and correctly positioned; the west coast of Norway correctly turns northeastward for a while, until it runs off the edge of the chart.

However, there are some inaccuracies that demand attention and will lead us to a surprising alternative interpretation. First, it seems questionable that Carignano and nobody else (for another two and a half centuries, as we shall see) should have had access to detailed data from an actual survey of Scandinavia. Second, if the details all appear so accurate, then to rule out coincidence one must pay particular attention to any outstanding inaccuracies. Such an inaccuracy exists at the lower end of the would-be Scandinavia. The coast shows two indentations instead of the one junction of Skagerrak and Kattegat (the Vik), or three promontories instead of the two bulges Scandinavia actually has. Such general information was known in Scandinavia and appeared properly in the Hereford Map (Figure 18).

I suggest that the apparent likeness of this map to the actual Scandinavia could be completely accidental. The possibility of accidental coastline similarities can cut both ways. Its actual ultimate source could be from Eskimo information about an Eskimo land, such as a map of that Eskimo land. Carignano's receipt of such a map prototype or description would naturally have been through Scandinavian sources, and he naturally would have taken it as a map of Scandinavia, under the provenance paradigm. This misunderstanding would be the precursor in the fifteenth century to a multitude of shapes identifiable as American in the new type of charts, which would culminate in the maps of Clavus, Behaim, et al., as discussed in Part I. Let us compare Carignano's outline with Greenland's nearest neighbor in Eskimo territory, Baffin Island (see front map). When the two outlines are just slightly rotated into similar alignments, one sees that the three promontories are correct for Baffin Island rather than Scandinavia. Carignano even shows an extra prominence on the back side in the exact position of Foxe Peninsula (see Figure 7). (The island in the Baltic could represent the actual Scandinavian island of Gotland.)

Such an occurrence of Baffin Island would be the next logical step, as the Greenlanders continued to pursue the question of Eskimo origins and the explorations started by the priests Arnold and Harold in 1266. The Thules had migrated into Baffin Island as well as northern Greenland.

Circa 1320: Vesconte / Sanudo / Paolino. A series of maps were drawn either by several different people working under the same master or by Petrus Vesconte himself while his thoughts evolved. Many of the maps were published bound into copies of a book by Marino Sanudo, and most of the maps differ slightly from one another. The maps were done in the style of the disk map school, but many of the coastlines seem to have characteristics similar to mariners' or portolan charts.

The first map (Figure 20), known to be a product of Vesconte himself, shows Scandinavia as a near-island connected to the mainland by a narrow isthmus. Kretschmer has shown that in fact Scandinavia was originally drawn on the map as an island; the isthmus was added after the rest of the map was finished.

The second example shown (Figure 21), from the Oxford copy of Sanudo's book, shows evidence of more compromise. The isthmus connection to the mainland is naturalistic and integrated but somewhat wider than Saxo's description (page 94) would indicate. The relative orientation of the Scandinavian countries and the Danish peninsula is quite accurate, but one wonders about the exaggerated prominences that appear on the west coast instead of the south. There are two major inlets in the north and one minor inlet in the south, just right for the west coast of Alaska. Other versions not reproduced here such as a Brussels manuscript show intermediate variations.[7]

The third example shown (Figure 22), from the Paris manuscript of Sanudo's book, has had its authorship assigned to Venetian Minorite, Fra Paolino.[8] It shows an apparent Danish peninsula that is bloated to a size larger than Great Britain. Beyond this, two giant islands lie amid several smaller ones. It is possible to argue that all these represent the islands between the Danish peninsula and Sweden, the whole Danish scale being mistakenly exaggerated. However, some nearby inscriptions suggest a possible reason for the exaggeration and an alternate identification of the islands as the Arctic Archipelago.[9] The enlargement of the scale might simply indicate that these islands represent the Arctic Archipelago, not the Danish archipelago. The inscription on the right side of the larger island calls it "Region of Frigid Denmark." Indeed, note the inscription "Noiegia" (Norway) in the margin at the end of the would-be Danish peninsula. This suggests that the peninsula rep-

Figure 20. Map by Petrus Vesconte, 1320. Location: Vatican, Cod. Palat. Lat. 1320. Source: Kamal, *Monumenta Cartographica,* vol. 4, fasc. 1, fol. 1160.

resents the entirety of European Scandinavia and that the islands are beyond Norway.[*] A marginal gloss next to these islands lists the various Germanic regions, for the map's more southerly readers, and the first one mentioned is "Frigid Germany." It says that among these islands is the island of Gyrfalcons. We have seen that this refers to either Greenland or the Arctic Archipelago.

The inscription on the smaller of the two giant islands can be interpreted, "In this sea is the greatest superabundance of *alcuorum.*" In its narrowest

---

[*]One of these islands (the long narrow one) carries a label whose interpretation has been controversial, sometimes interpreted as "Yslandia" (Iceland) and sometimes as "Gotlandia."

Figure 21. Map by Vesconte or his studio, in the Oxford copy of a book by Mario Sanudo, ca. 1320. Location: Oxford. Source: Kamal, *Monumenta Cartographica,* vol. 4, fasc. 1, frontispiece.

meaning, *alces* refers to the European elk (American moose). Differences in migration patterns between Europe and the American Arctic lead us to an American interpretation. It is true that moose have been known on unusual occasions to swim from Sweden to islands in the Baltic. However, this does not seem to be the kind of information to merit recording by a world cartographer. More generally, *alces* meant any large number of the deer family, in particular caribou. And in the Arctic Archipelago a great superabundance of caribou, in fact hundreds of thousands, did fill the sea twice every year during their migrations between the islands and the mainland. This was a

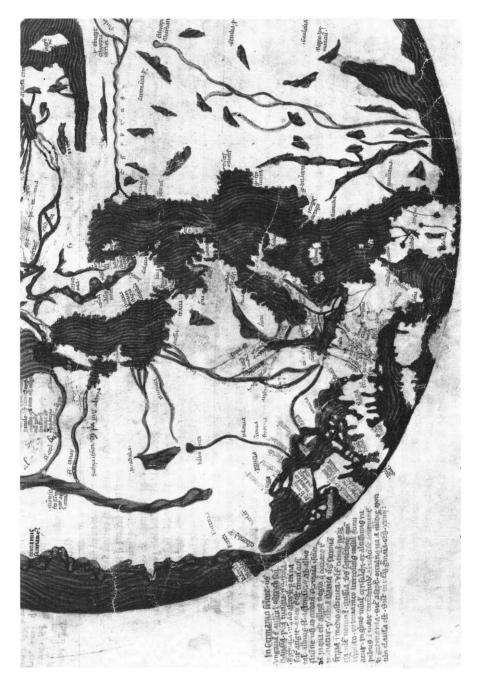

Figure 22. Map by Fra Paolino in the Paris manuscript of a book by Mario Sanudo, ca. 1320. Location and source: Paris, Bibliothèque Nationale de France, MS Latin 4939, fol. 9. Negative obtained from Bibliothèque Nationale de France, Paris.

spectacle described with wonder by every Arctic explorer, and it certainly would have merited a place on the map.*

Beyond the islands there is a long peninsula, pointing downward from the mainland. It is suggestive of the Alaskan Peninsula, and is the first appearance of such a peninsula in the cartographic record of Scandinavia. Beyond it we seen an early mention of the "Kareli Infideles"—presumably Thule Eskimos—who were to fascinate Clavus a century later.

1325: Dalorto-Dulcert. A Genoese-born mapmaker, Angelino Dalorto later moved to Majorca. There he changed the spelling of his name to Dulcert to conform to the local Catalan, a sister language of Spanish and French. Catalan cultural influence spread from Barcelona eastward under Aragonese political influence, into the Mediterranean lands as far as Greece and the Aegean islands. Cartography was an important part of Majorcan activities. Two of Dalorto's maps remain, and they are much alike.[10]

On the northwest detail of his map of 1325 (Figure 23) one sees a land mass labeled Noiueigia extending past the longitude of Great Britain. It originates, beyond the eastern edge of this detail, where Scandinavia properly does, east of Denmark. The scale of Scandinavia has been so enlarged that Norway reaches as far west as Scotland. Could this be another example of a new paradigm for accommodating to the fact that substantial land lies to the north and west? This idea is supported by his apparent inclusion of Iceland as an island labeled "Insula Tille," actually to the east of Scotland, with much of the would-be Scandinavian land to its west.

An inscription attached to the western part of Norway states that gyrfalcons are brought from there. Actually the white gyrfalcons came from Greenland and parts farther west. Another inscription relates to white bears, found in the American North but not in Scandinavia. Recall that we said there are entire graphic lineages of maps, not examined here, deriving from many of the coastline identifications we have made here. The same is true of animal species specific to America that appear in cartographic places like Scandinavia and Arctic Asia. They are all evidence of intellectual contact between southern European scholars and America.

---

*There has also been controversy about the perception of the word *alcuorum,* because of varying degrees of faithfulness in the reproductions. Some have read the word as *aletiorum,* meaning "herrings." Even a close examination of the original map with a magnifying glass might still leave some doubt, because of the similarity of *c* and *e*. The word *alcuorum* itself has two possible interpretations, one being "flying creatures." However, none of these alternate interpretations seems worthy of commemoration in a world map, while the one given seems quite natural.

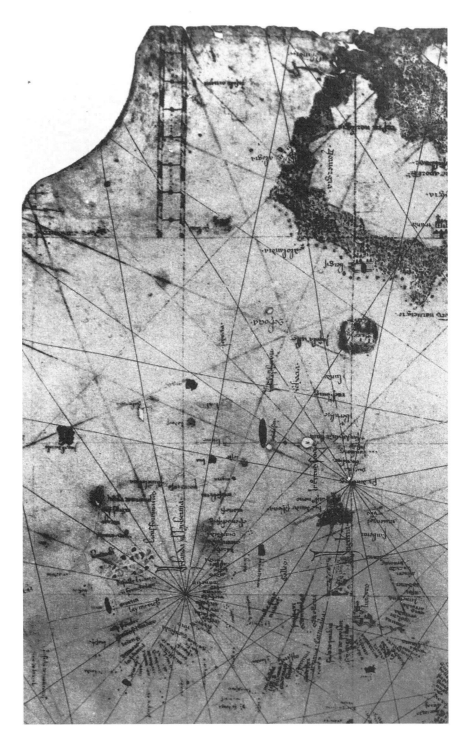

Figure 23. Northwest detail of the work of Angelino Dalorto, drawn in 1325. Location: Florence, Prince Corsini Collection. Source: Kamal, *Monumenta Cartographica,* vol. 4, fasc. 2, fol. 1197.

This map is the first to show the "mythical" island of Brasil, to the southwest of Ireland. Here in its initial Irish context this Gaelic name simply means "Isle of the Blessed."[11] This has exactly the same reference as the Latin *Insulae Fortunatae.**** As we follow the history of this island on later maps we will see that it seems to presage the strange bifurcated island in the Gulf of Greenland cited on Ruysch's map. That in turn seems to be a systematically distorted representation of Baffin Island.

Recall how new contact with the Thule Eskimos spurred resurrection of interest in the mythical Hyperboreans and Arimphians. The same thing happened with the Tartars and the Gog/Magog legends. Here it may be that new real geographical contacts resurrected interest in geographical myths. Dalorto's map shows, in the west coast of Ireland at the position of Galway Bay, a "Fortunate Lake" containing nineteen plus Fortunate Islands. As we follow the development of this idea we will see a possibility that it springs from the contemporary Norse-Eskimo contacts, whose thorough confusion with Irish legends and the Vinland Sagas may have transplanted the blessed places to the Irish coast.[†] Note that the number of islands, approximately nineteen, is the same as the number in the Inventio Fortunatae and in the Arctic Archipelago. In any case, their placement indicates that the old idea that the Fortunate Isles lay in the third climate off Africa had become flexible. Indeed, in a later (1339) map, Dalorto was the first to use the name *Canaria* for the islands off Africa formerly called the Fortunatae.[12]

Before 1333?: Kingigtorssuaq Runestone Cairns. Norsemen erected three rock towers forming an equilateral triangle off the west coast of Greenland on the tiny island of Kingigtorssuaq, at almost 73° north (see Figure 16). They set up rock-pile markers and beacons in many places but seldom inscribed any information on the rocks. The Kingigtorssuaq cairns, however, contained a stone inscribed with a runic message (Figure 24). It included stylistic factors from which some scholars have concluded that the cairns date from the early part of the fourteenth century. The explicit message reads: "Erling Sigvatsson, Bjarni Tordsson and Eindride Oddson on the Saturday before Rogation Day erected these cairns and cleared . . ." The ellipsis arises because the last part of the inscription is a jumble of apparently meaningless characters. This fact has combined with the occurrence of nonrunic characters elsewhere in

---

*To confuse matters, a second island in the Azores is given the name Brasil, but in that case it is likely not Gaelic. The country in the New World, Brazil, was named after its dye wood.
†One theory holds that the islands of Galway Bay inspired the Irish concepts of the Fortunate Isle of Brasil.

Figure 24. Norse runestone found in a cairn on the island of Kingigtorssuaq, far north of the Greenland settlements. Location, source, and permission to photograph: Copenhagen, Danish National Museum and Royal Danish Ministry of Foreign Affairs.

the inscription to lead several writers to believe that there is more information hidden in the inscription than has been explicitly understood. Norse rune carvers have been known to resort to cryptography in other inscriptions. One writer, Alf Mongé, believed that this stone contains a secret message that was encrypted by using the medieval Norse calendar as an encoding device. This theory has aroused much scholarly controversy, but it has been neither proved nor disproved in any conclusive way.

According to Mongé, the phrase containing the ellipsis should actually read, "and cleared an opening through the ice." In addition, the cryptogram supposedly gives the date as May 7, 1244, and the cryptographer as one Oirvar Valrslethn. If the cairns were actually erected as early in the year as May, there was certainly plenty of ice this far north through which an opening might be cleared. It is not apparent whether the ice was cleared from the ground to erect the cairns or from the sea to free the ships. The sawing of openings or docks through sea ice is well known to all Arctic sailors caught in desperate circumstances. This seems a more likely situation to merit recording in stone.[13]

In any case, the stone proves that Norsemen did go as far north as I postulated for identifying Kroksfjordarheidr, their starting point in the analysis of the letter of 1266 (page 105). If the date of 1244 is correct, then these cairns may have marked the previous farthest point north referred to in that letter.

1342: Gisle Oddsson's Annal. Sometime before 1673 the Icelandic bishop Gisle Oddsson compiled a chronology of Old Norse history. His entry for the year 1342 includes the statement that the "inhabitants of Greenland" had "spontaneously deserted the true Christian faith and religion, . . . repudiating all good morals and true virtues," and had turned to "*Americae populos,*" literally "the people of America." Oddsson was making good use of the new terminology, but when he was writing, the only people in Arctic America were the Eskimos. The meaning of his statement is not necessarily that the Greenlanders had converted to an Eskimo religion. It probably refers to some specific rebellion against the church's prohibition of commerce with the heathen Eskimos. Such an action would be quite likely if the royal supply ship had not shown up for a while. In fact, there is no record of any commerce between Greenland and Norway from 1327 until the ship bearing one Ivar Baardsson left Norway in 1341. It probably wintered in Iceland and arrived in Greenland during this incident effectively of rebellion.[14]

After 1341: Ivar Baardsson. The poor communication between Greenland and Norway affected religious institutional functioning, as well as daily life. In 1341 the Bishop of Bergen dispatched an emissary, a priest named Ivar Baardsson, to visit the Bishop of Greenland and presumably to obtain a first-hand account of conditions. Baardsson's report, known only in a third-person recounting, included a set of sailing directions for various localities. His description of the route to Greenland shows that ice conditions on the sea had worsened since the days of Erik the Red: "From Snaefellsnes in Iceland, which is shortest to Greenland, two days and two nights sail, due west is the course, and there lie Gunnbjorn's Skerries right in mid-channel between Greenland and Iceland. This was the old course, but now ice has come from the gulf of the sea to the northeast so near to the said skerries that none without danger to life can sail the old course and be heard of again."[15] There is a well established reason for this change in conditions. The cyclical variations of tidal strength are popularly known to depend on the relative positions of the sun and moon and the time of year (i.e., distance from the sun). In the nineteenth century Otto Pettersson developed a novel theory of tidal and climatic fluctuations that has been borne out by more recent research.[16] He calculated a grand cycle of extreme fluctuations of the strength of daily tides having a period of some 1800 years. Pettersson showed that the most recent tidal minimum had been in ca. A.D. 550 and the most recent maximum in ca. A.D. 1433. The heat budget of the Arctic is sensitively balanced by factors of sunshine and re-radiation, mixing of Arctic air with Temperate air and mixing of Arctic waters with other waters. The last factor is especially important because of the high heat-holding capacity of water and the mixing of the Gulf Stream with Arctic currents. These currents, as well as all other popularly familiar aspects of the sea, exist only in a surface "layer" of the ocean a thousand or so feet deep. Below this lies dense, highly saline water at 31° Fahrenheit. This cold subocean pervades all the seas as far as the Arctic and Antarctic "convergence," but, normally, little mixing takes place with upper waters. It can be seen that if mixing were to take place, the Arctic especially would be markedly affected by the loss of its slight heat to the still colder subocean. Pettersson discovered a mechanism whereby mixing can and does occur.

The tides create an advancing wave on the surface of the sea, and they also create a smoothly advancing internal wave on the "surface" of the cold subocean. This smooth wave advances in synchrony with the surface tide and does not cause any mixing with upper waters. But, when this underwater wave encounters a submarine mountain or continental slope, turbulence is created and mixing does occur. Conditions are favorable for this tidal mixing in the straits and islands of the Arctic.

One common misunderstanding of the theory has led to its misapplication by researchers and also to failure by laymen to differentiate between sea ice and land ice. While other studies do suggest actual climatic changes, this theory does *not* call for a change in atmospheric or oceanic temperature, or for a change in snowfall rates. It only calls for a change in heat *content* of the polar surface water sufficient to account for a reasonable fraction of the heat of fusion of sea ice. Thus, icebergs calved by the Greenland glaciers have always had to be treated with the same respect and the Greenland ice cap has always been the same.

Baardsson stayed in Greenland for several years and is not mentioned again in Norway until 1364. He served as steward of the bishop's residence at Gardar in the Eastern Settlement (near Julianehab—see front map). During this time, an event took place that confirms Gisle Oddsson's entry above for 1342: Ivar Baardsson was among those chosen by the law assembly to expel the Skraelings from the Western Settlement (near Godthab—see front map).

Nineteenth-century scholars took this reference to expulsion of the Skraelings to indicate hostilities, but that a visiting priest would have been sent to fight off invading Eskimos is untenable. Even the suggestion of an Eskimo invasion is untenable. While individual Eskimos protected themselves with ferocity, they seldom banded together for joint action. There is still a school of writers who posit a hostile Eskimo-Norse interaction, even explaining the ultimate disappearance of the Norse settlements this way. Admittedly, a couple of incidents of hostilities were documented from each side, but they have the ring of escalation, based on bravado, from trivial accidents. The hostility thesis is surely just as faulty as that of medieval writers in the British Isles who thought the Norsemen capable of nothing but pillaging. The same kind of thinking about "barbarians" may have created both of these attitudes. More likely perhaps, the group with Ivar Baardsson had an objective to enforce the church's ban on commerce with the Eskimos. They wanted to expel them in an ideological and economic sense. This explanation would be consistent with Gisle Oddsson's report, while making war would not; and it would suggest that the Western Settlement, three hundred to four hundred miles north of the Eastern Settlement, was already highly dependent upon Eskimo techniques of coping with life in the absence of European necessities. Indeed, by then seal hunting had supplanted dairy farming as the mainstay of life, and more walrus-based products were found in the Western Settlement than was typical at Eastern Settlement farms.[17]

The recounting of Baardsson's report continues: "And when they came there they found no man, either Christian or heathen, but some wild cattle and sheep, and ate of the wild cattle, and took as much as the ships could carry and sailed with it home." Note that the settlement had *not* been overrun by Eskimos. Nansen has pointed out that domestic cattle could not possibly survive the Greenland winter in the wild state. Thus, the vacating of the Western Settlement (or that part of it visited by Baardsson) could only have been temporary, for the summer if at all. The visitors may have been deceived by the entirety of a homestead, or several homesteads, packing up and temporarily moving elsewhere (e.g., to the northern hunting grounds for a summer period). This suggests to what degree the Western Settlement might have become independent of European necessities. The theft that the putatively misunderstanding clerics from the Eastern Settlement engaged in would have accelerated the adaptation process. This may be one step toward explaining why Hans Egede, who rediscovered the Greenland settlements in 1723, found relatively few artifacts in the ruins.

Kirsten Seaver has promoted another somewhat plausible scenario, theorizing that Ivar Baardsson was an ecclesiastical tax collector.[18] The western settlers had been seriously derelict in their tithe payments. She thinks they may well have fled into the interior when they saw Baardsson's fleet approach.

Beyond Baardsson's description of sailing directions for the places around the settlements and hunting grounds, he mentioned the following: "Item: there lies in the north, farther from the Western Settlement, a great mountain that is called *Himinrathsfjall,* and farther than to this mountain must no man sail, if he would preserve his life from the many whirlpools which there lie round all the ocean." This description fits very well the situation of the high glacier-covered peaks of Bylot Island at the north end of Baffin. This is where, I conjecture, the voyage of 1267 went (page 105), and beyond this lies the entrance to the whirlpool-like Indrawing Channels of the upcoming Inventio Fortunatae.

The visit of Ivar Baardsson is the last recorded evidence of anyone's having tried to contact the Western Settlement people.

SUMMARY

Perhaps the most important event in the entire history of Norse Greenland was its rediscovery of the Eskimos in the 1260s. At that point Norse society began evolving rapidly away from European society and towards the Eskimo culture. Many Norsemen traveled into Eskimo territory and perhaps abandoned their own permanent settlements. Their contacts with Eskimos led to textual descriptions in various Northern documents and may have led to European maps that, by means of the provenance paradigm, recorded good likenesses of the Eskimo lands from Baffin Island to the Arctic Archipelago and even to Alaska.

Europe seemed little interested in these matters, however, as it was making new discoveries of its own. The seeds of the Renaissance were being sown. The glorious stories of the Far East greatly overshadowed those of a wretched people from the West who might lead good Christians astray. The political forces that had drawn the attentions of Europe and Greenland toward one another for nearly a century had now dissipated.

UNCOVERING AN AMERICA

CHAPTER 9

LATE GREENLAND–BASED EXPLORATION

THE NORSEMEN MIGHT HAVE ENCOUNTERED AMERICA TWICE—once in Leif Erikson's time and again in the time of Ivar Baardsson. It will be difficult to document the geographical history of Greenland beyond the point when Europe stopped recording it officially, but there do exist isolated documentary suggestions that the Norsemen and some European visitors actively began exploring westward from Greenland. Although the accuracy of some of the documents in this chapter is controversial, even to the point of having been debunked by previous established authorities, recent scholarship on them suggests they are worth reexamining. Others of the documents are well established.

1347: Markland Voyage. The *Icelandic Annals* have the following entry for the year 1347: "A ship came from Greenland to Straumsfjord [in Iceland]; it had sailed to Markland, but later it was driven in here over the sea. There were eighteen men in the crew." Markland voyages, like Vinland voyages, had not been mentioned after the Erikson family. This may indicate a rediscovery of whatever was thought to be Markland. However, sources other than the *Icelandic Annals* emphasize not the Markland aspect of this voyage but that it had been shipwrecked. Thus, one cannot be sure which of these aspects was considered in 1347 to be newsworthy.[1]

1351: Medici (Laurentian) Marine Atlas. This collection of mariners' charts includes coverage of northwestern Europe resembling that in the Dalorto map. The depiction of Scandinavia (Figure 25) is even more extreme than in Dalorto, with the west coast reaching beyond Ireland. The portion around Denmark correctly shows the Skagerrak and Kattegat, while the expanded

Figure 25. The Medici (Laurentian) marine chart, 1351. Location: Florence, Biblioteca Medicea Laurenziana, Gad. Rel. 9. Source: Kamal, *Monumenta Cartographica*, vol. 4, fasc. 2, fol. 1246.

western portion is suggestive of the three prominences of Baffin Island as in Carignano.[2]

1363: **Inventio Fortunatae.** This work is covered in detail earlier in this book. It rather likely contains a detailed description of the Canadian Arctic Archipelago. Most of our information about its contents comes indirectly from an astrolabe-carrying priest whose report to the Norwegian king was relayed by one Jacob Cnoyen.[3] Some writers have speculated that, based on timing and date considerations, this Norwegian priest was none other than Ivar Baardsson.[4]

The meaning of the title seems to confirm the by-then established tradition of looking for the Insulae Fortunatae in the direction of the Eskimos and Vinland "the Good." The few explicit references that exist spell the title in varying ways which permit different interpretations of its meaning.[5] Las Casas's *Inventio Fortunata* would mean "a blessed (or fortunate) discovery," while the writer of a marginal note in Mercator's letter suggests *Inventio Fortunae,* meaning "Fortune's discovery." The grammatical parallel with *Insulae Fortunatae* (Isles of the Blessed) leads me to accept Cnoyen's spelling *Inventio Fortunatae.* This means "discovery of the blessed," reflecting a functional shift of the adjective *blessed* into a collective noun, *the blessed.* Perhaps the author was claiming that the Eskimos were the Blessed, the Arimphians. While such a misunderstanding is possible at a distance, his personal contact must have engendered the earliest "noble savage" philosophy. Nevertheless, this was consistent with the legendary characterization of the Hyperboreans. It would seem that in the mid-fourteenth century there was increased scholarly concern about just where the Insulae Fortunatae were. Starting at least with Dalorto/Dulcert's Fortunate Lake in Ireland, the concepts of Vinland and the Hyperboreans and Arimphians all fused into a belief that the Isles of the Blessed were in the North or Northwest.

That part of the contents of the *Inventio Fortunatae* that has been preserved for us does not deal with Vinland, but evidently with the Arctic Archipelago. The author mentions twice the discovery of grounded ships' balks and hewn planks in this area and concludes "with certainty" that people had been there before. These could have been Eskimos or earlier Norsemen from the Western Settlement. At two different places in his narrative the author refers to an island that sounds very much like Victoria Island. He describes the island west of the westernmost channel as the "best and healthiest land in all the North," and in another reference, which sounds like the same place, he says, "there lies a beautiful open land, which is uninhabited; there many beautiful . . . [manuscript burnt]." One might imagine to complete this sentence with the phrase "white falcons are caught."

Recall the inference in the analysis of the Yale Vinland Map that the *Inventio* contained systematic data about the same island that Dalorto called the Isle of Brasil. If correct, this suggests that the *Inventio* was not just a report of one man's journey but a general study of information on the Isles of the Blessed. Various writers have made speculations about the authorship of the *Inventio,* but none of them is without flaws. A popular theory that it was one Nicholas of Lynn is based on nothing but wishful thinking. Ferdinand Columbus cleverly referred to him as Juventus Fortunatus. Whoever the author was, it is likely that while in Greenland he worked closely with Ivar Baardsson.[6]

Circa 1372: Antonio Zeno's Fishermen. In the middle 1500s the New World was becoming well known throughout the Old World, and interest in possible pre-Columbian knowledge of those lands was mounting. In this environment, the Venetian nobleman Nicolo Zeno recalled an unpublished ancient family manuscript which he, as a child, had torn to bits.[7] He realized that it had contained invaluable information on the subject of just such exploration by his ancestors. He claimed to have tried to reconstruct the presumed content of the manuscript by patching together quotations from existing old family letters. In 1558 he published a book with the lengthy title *The Discovery of the Islands of Frisland, Eslanda, Estotilanda and Icaria: Made by Two Brothers of the Zeno Family: Viz. Messire Nicolo, the Chevalier, and Messire Antonio; with a Map of the said Islands.*

Unfortunately, the younger Nicolo, great-great-great-grandson of the Antonio in the title, was ill-equipped both educationally and critically for his undertaking. Later scholars struggled to various extremes with its contents. Some used the map (Figure 67) for piloting actual voyages. Frederick Lucas labeled the entire work "one of the most ingenious, most successful, and most enduring literary impostures which has ever gulled a confiding public." The question of authenticity is still controversial, but it is examined with some favor in the entry for Nicolo the younger in 1558 (page 277).

We examine here an extract from the period just after the older Nicolo's death, around 1398, in which the narrator is Antonio Zeno speaking from somewhere in the North Sea.

> Six and twenty years ago [i.e., ca. 1372] four fishing boats put out to sea, and, encountering a heavy storm, were driven over the sea in utter helplessness for many

days; when at length, the tempest abating, they discovered an island called Estotiland, lying more than a thousand miles to the westward from here. One of the boats was wrecked, and six men that were in it were taken by the inhabitants, and brought into a fair and populous city, where the king* of the place sent for many interpreters, but there were none could be found that understood the language of the fishermen, except one that spoke Latin, and who had also been cast by chance upon the same island. On behalf of the king he asked them who they were and where they came from; and when he reported their answer, the king desired that they should remain in the country. Accordingly, as they could do no otherwise, they obeyed his commandment, and remained five years on the island, and learned the language. One of them in particular visited different parts of the island and reports that it is a very rich country, abounding in all good things. It is a little smaller than Iceland, but more fertile; in the middle of it is a very high mountain, in which rise four rivers which water the whole country.

The inhabitants are intelligent people, and possess all the arts like ourselves; and it is believed that in times past they have had intercourse with our people, for he said that he saw Latin books in the king's library, which they at this present time do not understand. They have their own language and letters. They have all kinds of metals, but especially they abound with gold [copper?]. Their foreign intercourse is with Greenland, whence they import furs, brimstone and pitch. . . . They sow corn and make beer, which is a kind of drink that northern people take as we do wine. They have woods of immense extent. They make their buildings with walls, and there are many towns and villages. They make small boats and sail them, but they have not the lodestone, nor do they know the north by the compass. For this reason these fishermen were held in great estimation, insomuch that the king sent them with twelve boats to the southward to a country which they call Drogeo.

The inhabitants of Estotiland, a beer-drinking northern people who had their own alphabet and traded with Greenland, seem likely to have been Norsemen. Some scholars have concluded from the map and textual description that Estotiland must have been the island of Newfoundland. In accord with the sailing directions in the narrative, this lies due west of the North Sea by some 1,800 miles.

The narrative goes on to say that the boats heading for Drogeo encoun-

---

*Words such as *city* and *king* must be taken with reservation; they were probably Venetian interpretations of lesser entities. Only a few decades later there was an officially appointed king of the Canary Islands.

tered bad weather and missed their target, ending up somewhere to its south, in captivity. Robert Fuson has analyzed evidence suggesting that Drogeo was Nova Scotia.[8] The country to the south of Drogeo, "a great and populous country, very rich in gold," is then examined in some detail. The description includes references to the natives sounding very much like the North American Indians. The fishermen saved their hides by their novel ability to make fishnets. (There is no known archaeological evidence that the Beothuks, at least, ever made nets on their own.) Concerning the southerly natives, the Zeno account says:

> They are very fierce and have deadly fights among each other, and eat one another's flesh. They have chieftains and certain laws among themselves, but differing in the different tribes. The further you go southwestward, however, the more refinement you meet with, because the climate is more temperate, and accordingly there they have cities and temples dedicated to their idols, in which they sacrifice men and afterwards eat them. In those parts they have some knowledge of the use of gold and silver.

The stranded fishermen, according to the rest of the story, lived thirteen years among the natives south of Drogeo before returning thence, then to Estotiland, and ultimately homeward with their story.

The map is even more controversial than the narrative and is certainly not yet fully understood. There will be occasion to discuss this subject further below.

1375–1380: Catalan Fragment in Istanbul. This fragment of a map (Figure 26) is by an unknown member of the Catalan school of cartography. It is dated approximately by its similarities to a complete Catalan atlas of 1375 (Paris). The fragment is kept in the Topkapi Library, in Istanbul. The inscriptions are faded and photograph poorly, but we will rely on written reports by Marcel Destombes from inspections of the original. Right at the North Pole is a bay identified simultaneously as the White Sea and the Northern Caspian. Several other maps, including those of Petrus Vesconte and Fra Paolino, contained such a doubling of the Caspian Sea as well as other features like "Albania." The sea contains two islands, labeled "Salmos" and "Naron." Salmos is a classical Greek city described by Herodotus as an island city he called Samos. Naron was in northwestern Spain.[9]

Such doubling and northern relocation has never been adequately explained, and this writer will not attempt to explain it. But the attribution of

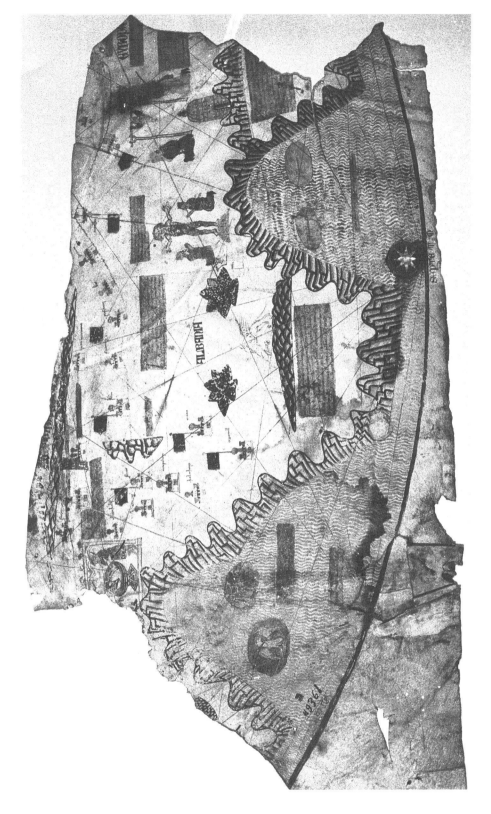

Figure 26. Catalan map fragment in Istanbul, ca. 1380. Location and source: Istanbul, Topkapi Saray Kutuphane, H.1828. Courtesy of Topkapi Palace Museum.

it simply to ignorance is no explanation at all. We have previously seen, as with the Arimphians and Hyperboreans, that when classic material appears in an out-of-context location it may represent an attempt by contemporary thinkers to interpret some anomalous real phenomenon or people in classic terms. One such phenomenon that might conceivably apply here is new information about the Arctic Archipelago. In a later map (Figure 40) we will see Naron take on the distinctive shape of Victoria Island. To the east is another bay containing two islands; they bear the insignia of a white falcon and an inscription about many beautiful falcons and gyrfalcons in the land of Marco Polo's Grand Khan. These islands could represent Greenland and Baffin Island, perhaps by then acknowledged to be on the eastern rather than the western end of a continental land mass. These sentences would have been the rankest of speculation without the interpretation of the *Inventio Fortunatae* given in Chapter 3. But the closeness in time of the Catalan fragment to the *Inventio* supports the idea. The stories of Marco Polo naturally would have assisted in such a transition, even though his mention of falcons is very minor and off-handed.

1385–1387: Björn Einarsson. Björn was a widely traveled Icelandic trader who even made a pilgrimage to Jerusalem. In 1385, while on his way from Europe to Iceland with four ships, he was driven westward by a severe storm and sought refuge in the Eastern Settlement of Greenland. The Greenlanders did their best to feed the hungry crews and trade with them, but the sailors experienced great shortages. The situation was saved when Björn encountered some Eskimos:

> He came to the assistance of two Skraelings, a young brother and sister, caught on a tidal skerry. They swore fidelity to him, and from that time he was never short of food; for they were skilled in all kinds of hunting, whatever he wished or needed. What the Skraeling girl liked best was when Solveig (the mistress of the house) allowed her to carry and play with her boy who had recently been born. She also wanted to have a linen hood like the mistress, but made it for herself of whale's gut. They killed themselves by throwing themselves into the sea from cliffs after the ships, when they were not allowed to sail with the franklin Björn, their beloved master, to Iceland.[10]

The most historically striking aspect of this story is its demonstration of a thoroughly changed attitude in the Eastern Settlement toward contact with the Eskimos. Norsemen now not only permitted but welcomed them into household contact and accepted them as saviors from an otherwise intolerable economic situation. This economic situation is markedly different from that of the early days when Leif and Erik put up Karlsefni's two crews in their own household for the winter, and it suggests a decline in both population and economic well-being of the Eastern Settlement.

Circa 1395: Nicolo Zeno's Greenland. Sometime after his brother Antonio joined him in the North Sea, Nicolo, according to the story, went on a journey to Greenland. He left us with a description of the conditions he saw there at a monastery situated by a hot spring whose location is unknown. (The only currently active hot spring in southwest Greenland is at Unartoq, which would have been in the Eastern Settlement, where a cloister actually did exist.[11]) The description, if true, shows that by that time the monks more than just condoned commerce with the Eskimos. In fact, the monks had absorbed some of the Eskimo culture. However, the narrator was most taken with the thermal springs, which he found to be used much as springs in Iceland were used.

> There is a spring of hot water there with which they heat both the church of the monastery and the chambers of the Friars, and the water comes up into the kitchen so boiling hot, that they use no other fire to dress their victuals. They also put their bread into brass pots without any water, and it is baked the same as if it were in a hot oven. They have also small gardens covered over in winter time, which being watered with this water, are protected against the effect of the snow and cold, which in those parts, being situated far under the pole, are very severe, and by this means they produce flowers as in other temperate countries in their seasons, so that the rude and savage people of those parts, seeing these supernatural effects take those friars for Gods, and bring them many presents, such as fowls, meat, and other things, holding them as Lords in the greatest reverence and respect. When the frost and snow are very great, the friars heat their houses in the manner described, and by letting in the water or opening the windows, they can in an instant temper the heat and cold of an apartment at their pleasure. In the buildings of the monastery they use no other material than that which is supplied to them by the fire; for they take the burning stones that are cast out like cinders from the fiery mouth of the hill, and when they are at their hottest they throw water on them to dissolve them,* so that they become an excellent white lime

---

*This description does not make sense as an erupting volcano, as previous writers identified it and which they felt constrained to locate outside Greenland. But it may describe some kind of erupting mud pit or other source of unusual building materials common at hot springs.

which is extremely tenacious, and when used in building never decays. These clinkers when cold are very serviceable in place of stones for making walls and arches; for when once chilled they will never give or break unless they be cut with some iron tool, and arches built of them are so light that they need no strong support, and are everlasting in their beauty and consistency.* By means of these great advantages these good friars have constructed so many buildings and walls that it is a curiosity to witness. The roofs of their houses are for the most part made in the following manner: first, they raise up the wall to its full height; then they make it incline inward by little and little, in form of an arch, so that in the middle it [admittedly] forms an excellent passage for the rain. But in those parts they are not much threatened with rain, because the pole, as I have said, is extremely cold, and when the first snow is fallen it does not thaw again for nine months, which is the duration of their winter. They live on wild fowl and fish; for, where the warm water falls into the sea, there is a large and wide harbor, which, from the heat of the boiling water, never freezes all the winter, and the consequence is that there is such an attraction for sea-fowl and fish that they are caught in unlimited quantity, and prove the support of a large population in the neighborhood, which thus finds abundant occupation in building and in catching birds and fish, and in a thousand other necessary occupations about the monastery.

Their houses are built about the hill on every side, round in form, and twenty-five feet broad, and narrower and narrower towards the top, having at the summit a little hole, through which the air and light come into the house; and the ground below is so warm, that those within feel no cold at all. Hither, in summer time, come many vessels from the Cape above Norway, and from Trondheim, and bring the Friars all sorts of comforts, taking in exchange fish, which they dry in the sun or by freezing, and skins of different kinds of animals. By this means they obtain wood for burning, and admirably carved timber for building, and grain, and cloth for clothes.† For all the countries round them are only too glad to traffic with them for the two articles just mentioned; and thus, without any trouble or expense, they have all they want. To this monastery resort Friars from Norway, Sweden and other countries, but the greater part come from Iceland. There are continually in the harbor a number of vessels detained by the sea being frozen, and waiting for the next season to melt the ice. The fishermen's boats are made like a weaver's shuttle. They take the skins of marine animals, and fashion them with the bones of the self-same animal, and, sewing them together and

---

*Muds in some thermal areas are highly infused with gas bubbles and when hardened would have the qualities described.

†The Greenlanders usually wove their own clothes, but, as Nörlund has shown, they were always eager for the latest fashions from Europe.

doubling them over, they make them so sound and substantial that it is wonder-
ful to see how, in bad weather, they will shut themselves close inside and expose
themselves to the sea and the wind without the slightest fear of coming to mischief. If they happen to be driven on any rocks, they can stand a good many
bumps without receiving any injury. In the bottom of the boats they have a kind
of sleeve, which is tied fast in the middle, and when any water comes into the
boat, they put it into one half of the sleeve, then closing it above with two pieces
of wood and opening the band underneath, they drive the water out; and this
they do as often as they have occasion, without any trouble or danger whatever.[12]

The modern reader will be impressed with the clarity of Zeno's reportage,
and especially with his precise description of an Eskimo kayak. The important thing to note, however, is that these kayaks were not being used by Eskimos but by the fishermen from the monastery. Evidently the fraternization
between the Norsemen and the Eskimos had proceeded to the point where
the monks themselves had realized the value of Eskimo techniques. But this
is just the beginning, for the description of the round domed stone houses
is a perfect description of the Thule snow igloos, complete with vent. Here,
however, a more durable building material was acquired from the thermal
area. The houses were built "about the hill on every side," and there were "so
many buildings and walls that it is a curiosity to witness." This is a further
perfect description of a typical Thule village, with multiple dwelling units
interconnected by arch-walled passageways, albeit of different materials. Yet,
the inhabitants of this village were not Eskimos but European Christians who
traded with the "rude and savage people of those parts" for fowls, meat, and
other things and received guests from Venice. If authentic (see page 279), they
clinch two beliefs about the Greenlanders. The belief that the Greenlanders
were turning to the Eskimos for aid is certainly confirmed, while the belief
that they were turning to heathenism is certainly denied.

SUMMARY

The Greenlanders had become neo-Vinlanders. Even ignoring the controversial documents, the *Inventio Fortunatae* suggests strongly that the early Renaissance Norsemen and their visitors had explored westward from Greenland. The controversial material raises the possibility that they also turned
southward. In any case, the fixed settlements in Greenland were no longer
tenable, and the classical European culture there was coming to an end. But
that is not to say that the intellectual effect of the Norse contacts with American on the European mind was at an end.

CHAPTER 10

FOUNDATIONS OF EUROPEAN

MISUNDERSTANDINGS

THE EARLY FIFTEENTH CENTURY IN EUROPE SAW INTELLECTUAL FERMENT on a variety of geographical subjects. It used to be fashionable to explain this as merely another facet of the growing Renaissance, but contemporary scholars have realized that ascribing every intellectual movement to the Renaissance is no explanation at all. Today it is possible to see the Renaissance as the logical confluence of many movements, all of which were firmly rooted in the activities of preceding ages. The multiethnic contact at this time, including European interaction with Moslems and Jews, provided much stimulus to broaden horizons. At least part of the movement in geography, as this chapter will argue, may have been based on the Norse Greenlanders' discovery of the American eastern seaboard, also a discovery involving multiethnic interchange.

Circa 1406: First Latin Ptolemy. Following Ptolemy's death in 168 A.D., the science of astronomy and all other sciences had deteriorated. Soon practically the only people who knew about Ptolemy's work were Greek and Arab scribes recopying old books. At the dawn of the fifteenth century a Byzantine teacher of Greek in Italy, Emanuel Chrysoloras, needed a suitable exercise for one of his advanced students, the Florentine Jacopo d'Angiolo. The outcome, done ca. 1406, was the first translation of the Ptolemaic text into Latin. For the first time the Western world could see Ptolemy's theory.

Ptolemy's scientific geography sprang from the detailed consideration, by his near contemporary Marinus of Tyre, of one simple fact: because of the sphericity of the earth, one must differentiate between chorography, the mapping of a single region or country, and geography, the mapping of the entire world. In chorography, the earth may be treated as flat, and this will cause only minor distortion of the map. But in depicting the entire earth on a two-dimensional surface one cannot simply join together adjacent chorographic maps. That would not give a correct idea of what the round earth would look like to an observer off in space. Thus, Ptolemy developed a system of projections (both conical and spherical) grounded in the division of the globe into degrees of spherical latitude and longitude (which had originated centuries before for astronomical purposes). He showed how points should be mapped relative to this fixed framework rather than in relation to neighboring locations. As a byproduct, he introduced multiple reproductions of a map from tables of latitude and longitude instead of from tracing. This would introduce fewer progressive errors than are found when a map is recopied by necessarily imperfect tracing for several copy generations. Along with the text of this *Geographia* Ptolemy published an atlas of chorographic maps (unprojected) and a world map on the conical projection (see Figure 11 as a conceptual example).

The influence of the rediscovery of Ptolemy on Renaissance cartography was electric, but not immediately universal. Many either did not understand the mathematics of it or wished to avoid the requirements for precision. Nevertheless, most realized that Ptolemy's work contained a wealth of chorographic detail that could be used to fill in blank spaces on their world maps. In reading him for that purpose they gradually began absorbing the theory.[1]

Note, however, that people were not entirely ignorant of mathematical geography before this. For example, decades earlier, in 1363, the author of the *Inventio Fortunatae* knew the concept of measuring latitudes with an astrolabe, and his writing a book presumes a knowing audience. Lully's *L'Art de Navegar* had described use of the astrolabe nearly a century before. Chaucer's 1391 *Treatise on the Astrolabe* was a relative latecomer, but it demonstrates that this knowledge existed in nontechnical circles.

1410: Pierre d'Ailly. The extensive writings of Pierre d'Alliy, Bishop of Cambrai, in northern France, and later Cardinal, included not only theological works but many astronomical and geographical treatises. In 1410 during the initial wave of excitement over the rediscovery of Ptolemy, he wrote a book, *Imago Mundi* ("The Depiction of the World") containing a detailed theoretical discussion of "the Spheres," and the Earth in particular. In discussing the overall geographical characteristics of the Earth, especially the old theoretical northern boundary of the habitable climates, he says: "There is many a habitation beyond it, as for example Anglia, Scotia, Dacia, Norwegia, and many other countries the ultimate of them being according to some authorities Tyle. According to Aristotle and Averroes at the close of his *De Coelo et Mundo,* the end of the habitable earth on the east and the end of the habitable earth on the west are rather close; and in between is a scanty sea as to its breadth, although by land the way [from the West to the East via Asia] would be more than half the circumference of the Earth." This idea was not completely new to d'Ailly's learned readers, who also could have seen it in Roger Bacon (as d'Ailly did). But they might have been surprised to see him propound it so closely following his remarks of a few lines earlier, "Regarding the line which bounds the climates on the west there are few or no habitants except perhaps on some little islands, since there is a great sea there which is called the Ocean." The contradiction between "great sea" and "scanty sea" bothered d'Ailly, and he cited many highly respected authorities on both sides of the issue. Even though most authorities felt that the sea covered at least half of the globe, d'Ailly unscholastically entertained the theoretical arguments of the minority in favor of a relatively "scanty" ocean. He even advanced his own theoretical reasoning:

> There ought to be an abundance of water toward the poles of the earth because those regions are cold on account of their distance from the sun; and the cold condenses moisture. Therefore the water runs down from one pole toward the other into the body of the sea and spreads out between the confines of Spain and the beginning of India, of no great width, in such a way that the beginning of India can be beyond the middle of the equinoctal circle [the end of the hemisphere] and approach beneath [i.e., around] the earth quite close to the coast of Spain.

Just how scanty did d'Ailly think the ocean was? His text gives no quantification, but his Seventh Figure (Figure 27) gives a hint. This circular zone-map is not a representation of the classical mappenmundi depiction of a circular world continent, but rather is a modern hemispherical view of the entire globe. While d'Ailly did not depict any continental outlines on this figure, he did show some major rivers and waterways, including an outline of the Mediterranean Sea. A generally overlooked feature that did appear on many of the original manuscript copies of d'Ailly's work is an outline of "the Ocean." Along the northwestern edge of the hemisphere he placed the word *Oceanus* twice, and just inside this word he showed the coastline of western Europe. However, on the outside, apparently hanging in space, he also drew his idea of the far shore of the Ocean. How are we to interpret this quan-

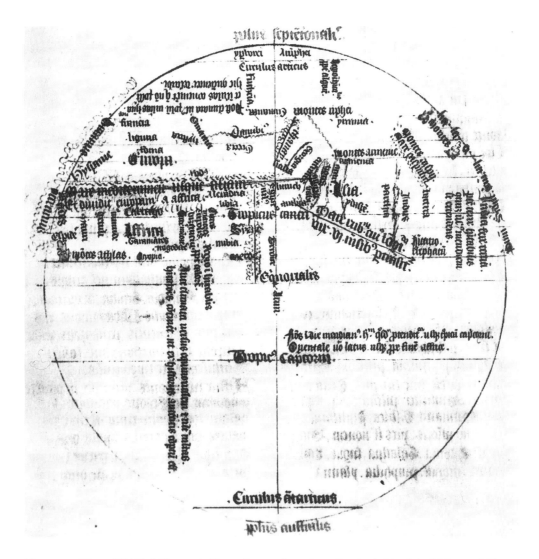

Figure 27. Pierre D'Ailly's "Seventh Figure," 1410. Location: Cambrai, Municipal Library, MS 954. Source: Edmond J. R. Buron *Ymago Mundi de Pierre D'Ailly* (Paris, 1930), vol. 2, p. 404. Courtesy of General Research Division, The New York Public Library, Astor, Lenox and Tilden Foundations.

tification of the oceanic width? Was it, as seems apparent on the figure, approximately equal to the width of the Mediterranean?

The answer may be contained in the very fact of the precision with which d'Ailly approached all his work. For him to have indicated the far shore vaguely as suspended indefinitely in space would have been out of character. More likely this was a device by which to suggest that the coastline actually lay on the back side of the globe, at a projected distance from the horizon equal to the distance it extended into space.[*] The far coastline would practically overlap the near coastline if one were looking through a transparent globe.[†] The quantitative interpretation of this device requires an appreciation of the principles of projection involved when one's line of sight is nearly tangent to the edge of the globe. D'Ailly was quite aware of those principles, as another work (his *First Treatise on Cosmography*), in which he described the technical procedures involved in constructing such a global figure, indicates: "Therefore this figure conforms to the principles of mechanics. Nevertheless, all the difficulty which occurs in the sketching of the habitable world on a plane would have been excluded in its drawing on a sphere." Imagine in your mind's eye the actual sphere of which d'Ailly's Seventh Figure is a projection, or endow this figure with depth. You will realize that the distance actually represented between the two shorelines is quite a bit greater than the apparent planar width, as it now includes the added depth of the globe. In fact, imagine facing an actual modern geographical globe in the same position as d'Ailly's figure, with the meridian line centered over Jerusalem. It can be seen that while the near coastline of d'Ailly's does coincide with the coast of Europe, the far coastline would correspond exactly with the coast of North America.[2]

1415: Albertin de Virga. Nothing is known about mapmaker Albertin de Virga except that he drew a map of the Mediterranean and Europe in 1409 and the world map being studied here around 1415 (Figure 28). His world map demonstrates an amazing curiosity about Scandinavia, which it shows as a giant subcontinent as big as Europe itself. This part of the map is enlarged in Figure 29. Neither the size nor the shape of this land could have been inspired by anything real in Scandinavia. The natural inclination of historical cartographers until now has been to dismiss as a fantasy all except the Baltic and North Sea areas, which are identifiable with portions of earlier maps. However, as we have seen in Part I, many curiosities in Scandinavia can be explained otherwise under the provenance paradigm. And the first of those misunderstandings we examined, Clavus's map of 1427, was not long to come in 1415. Our natural reaction by this point should be to look for some

---

[*]Just such a device was used in the early 1600s by the Renaissance scientist Mersenne. In his fifth book, on wind instruments, at Proposition 33, he depicts finger holes on the back side of cylindrical musical instruments as suspended in space, just like D'Ailly's Asian coastline.

[†]Many later printed editions do not show the far coastline in space but rather draw it almost superimposed on the near coastline. This confirms our interpretation of the device and confirms that contemporaries understood it.

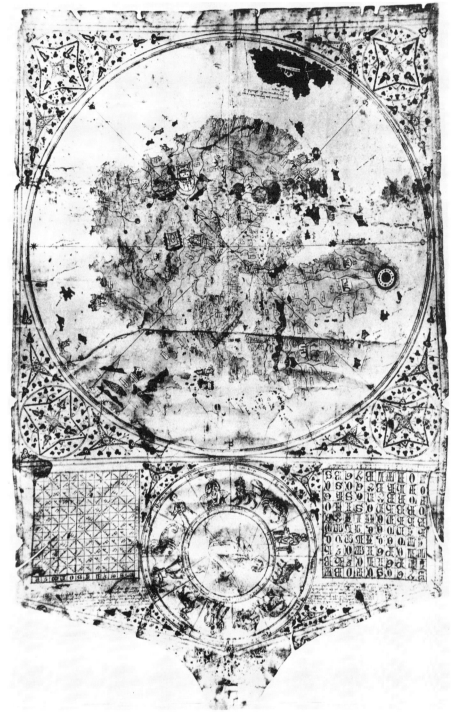

Figure 28. Albertin de Virga's 1415 world map. Location: Unknown. Source: Kamal, *Monumenta Cartographica*, vol. 4, fasc. 3, fol. 1377.

part of the New World whose transmission via the Greenlanders and Scandinavians might have reached de Virga and been misconstrued as Scandinavian land. This could have been forced by a reluctance to concede to notions such as d'Ailly's concerning the distance to Asia required by the one-ocean paradigm.[3]

The overall shape of de Virga's Scandinavian outline is as a large wedge shape or pie sector, attached at its apex to the Eurasian mainland just above Denmark. Except for the scale, this presages the shape of "Pilapaland" (alternately sometimes "Engroneland") in the upcoming maps derived from Claudius Clavus (Figure 47). Another later map, not shown here, that contains a similar feature is the Zeitz map of 1470.[4] Still other later maps will show a similar wedge shape in Asia under the one-ocean paradigm. Near de Virga's attachment point, just before the apex and directly north of Denmark, a thin dark line wiggles its way across from one shore to the other, with a larger dark spot midway. This is identified on the detailed view as a lake with short rivers flowing to each shore. I can hypothesize just one piece of land in the New World that fits these characteristics, and we know that information about it could have reached the Norsemen. That piece of land would be, perhaps no longer outragously, the entirety of the wedge-shaped Quebec-Labrador peninsula (see front map), flanked on one side by Hudson Bay and on the other by the St. Lawrence estuary.

At the apex of this Quebec peninsula are the Great Lakes, not quite but almost giving rise to rivers toward each shore. The lakes' regular exit is through the St. Lawrence. A slight elevation just north of the lakes, in the area labeled Nipigon, prevents their draining into rivers flowing to James Bay. However, there are smaller lakes on this divide, such as Long Lake, which do drain in both directions. The inclusion of these with the larger ones would seem plausible.

De Virga's inability to realize *which* continent the Quebec peninsula should be attached to is understandable. Nevertheless the scale on which he draws the land suggests an awareness of something of at least subcontinental proportions off in the ocean of Scandinavia. Recall the long thin peninsula in the Paolino Paris manuscript (page 125) that looked similar to the Alaska Peninsula. Here it seems to be separated from Scandinavia and attached at the far end of the new land. This would be appropriate for the Alaska Peninsula's description as lying at the extreme end of the continent.

A basic question raised by this interpretation of the map would be the following: while all the previous maps indicating traces of North American land could be attributed to Eskimo information, this one cannot completely.

While the Eskimos did enter Ungava and Labrador and went fairly far south, they never ventured so far south in Quebec or Ontario as the watershed divide or the Great Lakes. Nevertheless, the Norsemen are known to have had contact with American Indians as well as Eskimos. An Indian arrowhead was found in a Greenland burial, presumably the cause of the occupant's death. Presumably his body was brought back from somewhere south for burial in sacred ground.

The Indians of the Northeast were for the most part rather friendly when treated civilly, and one of the first questions, even in sign language, when meeting a new people civilly is, "Where do you people come from?" or "What's back there?" In 1603 Champlain made an early encounter with Indians in the St. Lawrence Valley who knew through trading contacts about geography up to Hudson Bay and down to the Atlantic seaboard, including the Great Lakes. American Indians *did* make hard copy maps, especially when asked to do so, and certain of their characteristic stylisms may be relevant in this picture by de Virga. Lakes are usually represented as circles, and complex structures (e.g., the Great Lakes) are often simplified.[5] So, it is not completely outrageous to suggest that this 1415 map could represent information about the Quebec-Labrador peninsula.

To the north of the suggested Quebec peninsula prototype are three islands (Figure 29) presumably derived from the Arctic Archipelago, bearing the names *Armana, Dicor* and *Singui*. While I have not been able to trace any origin for the names *Armana* and *Dicor,* the name *Singui* comes straight from Marco Polo's *Travels.* There the word *Singui* refers both to a province in eastern Asia and a city within the province. The province was described as economically important because of its vast production of salt, as well as fine rhubarb and ginger. The city itself, whose name Polo says means "City of the Earth," is situated near the river Kiang (Yangtze), "which is the largest river in the world," and was described as having a river port that made it a place of great commerce.

So, while we invoked the provenance paradigm to bring the Quebec-Labrador peninsula into Scandinavia, de Virga displayed an Asian outlook in Scandinavia. The significance of de Virga's apparent mixing of a western Norse geographical contact with Marco Polo's eastern Asia is more than academic, for he must have believed not only that it was theoretically possible to reach the Asian lands via Scandinavia but that in fact somebody had already done so. Herodotus had suggested that Phoenicians in antiquity might have gone around Africa to Asia, but this is using contemporary direct evidence to prove the reality of the "Grand Misunderstanding" engendered by

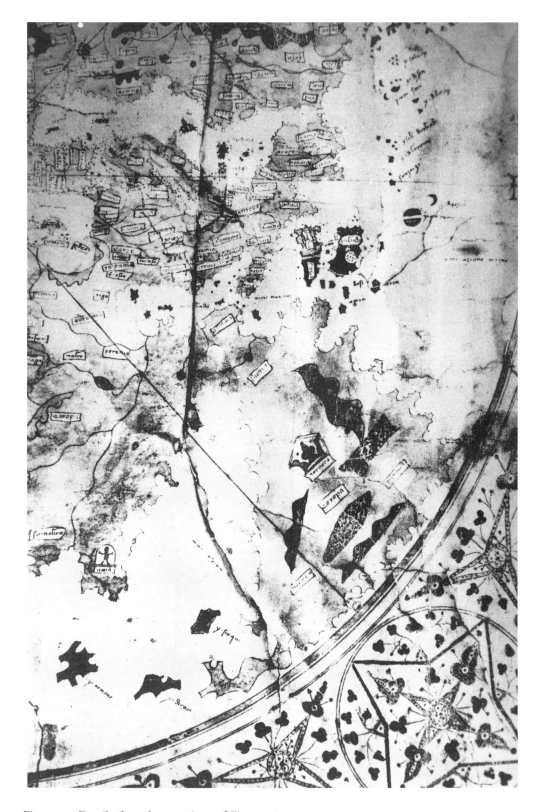

Figure 29. Detail of northern regions of Figure 28.

the simultaneous existence of the one-ocean paradigm and the provenance paradigm. De Virga's strict devotion to scientific fact in both of his known maps shows that this belief must have been grounded on convincing evidence.

His placing of Polo's names in places that were effectively in the western ocean highlights other appearances of Polo's place names in the West. For one, the *Gestae Arturi* cited by Cnoyen, which may have been contemporary with the 1360 *Inventio Fortunatae,* already used names from Marco Polo, such as the province of Bergi, putatively in connection with the Arctic Archipelago. For another possible occurrence of a Polo name in the West, note the island named Brasil, which first appeared in the western ocean in Dalorto's 1325 map (Figure 23), where we surmised it was Gaelic for Isle of the Blessed. But Marco Polo also described an island he encountered in the East (now called Java) which was famous for its Brazil wood. Indeed, a descriptive passage in Cnoyen's summary of the *Inventio Fortunatae,* not quoted here, mentions Brazil wood, and a marginal comment remarks on its importance.[6] It is difficult to see how de Virga reconciled his northerly insular Singui with Marco Polo's placement of Singui on the mainland.* Nevertheless, Singui's commerce reminds one of Clavus's reference to the market town or emporium of Ynesegh, also far north, at the tip of an island (page 45).

Perhaps the most fundamental significance of de Virga's Singui is that he placed it where he did without any explanatory scrolls or inscriptions. It has been fashionable for historians to assume that in the fifteenth century the concept of sailing westward to reach the East could be grasped only after lengthy explanation. De Virga seems to deny this, at least for his intended audience. The same is true for Cnoyen and the *Gestae Arturi.* This kind of thinking may also illuminate why the Yale Vinland map in particular was chosen to be attached to the *Speculum Historiale* and *Tartar Relation,* another description of eastern Asia.

1418: Sacking of Eastern Settlement by Pirates. An account was preserved indirectly in a letter written in 1448 from Pope Nicholas V to the bishops of Iceland. It implored them to come to the aid of the few survivors of the pirate attack. The account goes: "From adjacent heathen coasts [i.e., a hideout on Eskimo territory] the pirates came with a fleet, attacked the inhabitants of Greenland with a cruel assault, and so destroyed the fatherland and the sa-

---

*In fact, maps would soon relocate the Quebec-Labrador peninsula to eastern Asia, switching from the provenance paradigm to the one-ocean paradigm.

cred edifices with fire and sword that only nine parish churches were left in the whole island, and these are said to be the most remote, which they could not reach on account of the steep mountains. They carried the miserable inhabitants of both sexes as prisoners to their own country, especially those whom they regarded as strong and capable of bearing constant burdens of slavery, as was fitting for their tyranny."[7]

Some writers have speculated that these pirates were from England, which did have increased interest in the Scandinavian sphere at this time. This account is confirmed in amplified detail, in ancient Eskimo narratives by witnesses of the event, collected by Niels Egede. It is the *last record* directly showing any contact between Greenlanders and Europe. By this time Norway had been wracked by revolution, political intrigue, and foreign kings. The only recorded concern shown for Greenland was the ineffective plea of the Church in Rome for neighboring Iceland to lend spiritual aid. This incident is generally supposed to mark the end of the Norse occupation of American lands. However, we shall see that more than the ghosts of the Norsemen were still in Greenland. We shall see that succeeding documents hint at continuing activities of these inhabitants of isolated Greenland.

1422–1439: Duminicus Ducier. I have found photographs of the manuscripts in Figure 30 at the U.S. Library of Congress but have not been able to find their source. The pages are labeled "B," "C," and "D," so apparently part of the manuscript is missing (page "A"). The backs of the photographs are labeled "Gift, F. Rossi, 3N, '08," but the library was unable to trace this provenance.[8] One Marcian F. Rossi is known to have been the source of the so-called "Polo Maps" published in 1948.[9] These were part of a larger Rossi family collection, much of which was destroyed in the San Francisco fire of 1906.[10] Perhaps the photographs here record some of the collections earlier members. To my knowledge, images of this manuscript are here published for the first time. The manuscript is written by a Venetian scholar, one Ts. Visco. It describes the work of his French contemporary, Duminicus Ducier. Some think the calligraphic style does not match what would be expected for the era claimed, and it could be a later copy or, of course, a forgery, as some of the later Rossi Polo maps are suspected of being.

Ducier's mappemonde on page "B," claiming to have been drawn April 4, 1422, shows a continent labeled "Tullia Majore" in the western ocean, stretching from a connection in Scandinavia to a connection with Africa. This anticipates Björnbo's interpretation of the Norse idea that Vinland was connected with Africa and would be an example of the modern "inland At-

"B"

Figure 30. World maps by Duminicus Ducier, 1422 ("B" and "C") and a contemporary (1439) description by "Ts. Visco" ("D"). Location and source: Washington, D.C., Library of Congress, Map Division, Title Collection, *World, 1422–39, Ducier.*

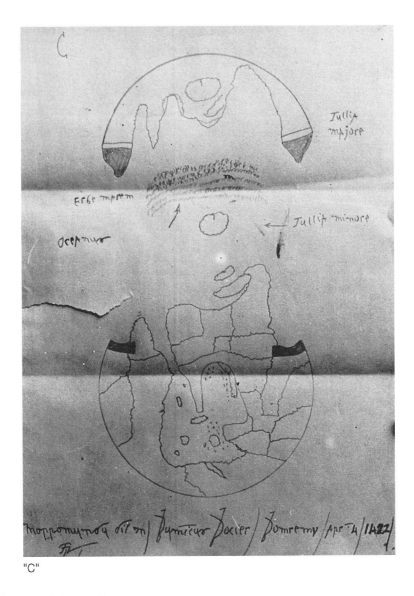

"C"

Figure 30. (*Continued*)

lantic" interpretation rather than the one-ocean paradigm. The place named
Tullia suggests a derivation from Pliny's Thule, the ultimate island. Alternatively, since it was reported by one Tullius Panfellius, it may have gotten its
name in the same grammatical way that Martin Waldseemüller gave America its name as "Land of Americus Vespucius." The text[*] says: "The form of

---

[*]A translation of the following by Benjamin B. Olshin and the author: "Furma dil Glubu prima
dil diluviu / facto da Duminicus Ducier / opera basata sur raporti di Tulliu Panfelliu oracolu di
Quintus Sestorius / cui scriptura dise che spiritu maliju afferoe la nave di Sestorius nil gulfu di
Biscania et la tiroe in hun habisso nil mare oceanus et poi por indulcenza dil oracolu fue mesa

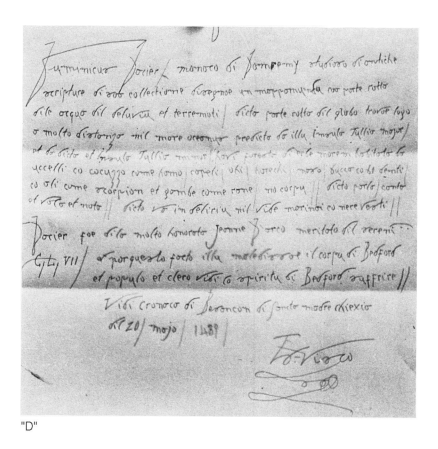

"D"

Figure 30. (*Continued*)

the world before the Deluge, made by Duminicus Ducier. The work based on reports of Tullius Panfellius, oracle of Quintus Sestorius, whose writings say that evil spirits seized the ship of Sestorius in the Bay of Biscay and drew it into an abbyss of the Ocean, and then, indulging the oracle, it was set adrift and carried into the coasts of the island Tullia Major, where the sailors detained themselves with the female aborigines. Sestorius and the oracle returned to Biscay. . . . / Padua, 7 J., 1426 / T.V." The Quintus Sestorius referred to might be thought to be the Roman general Quintus Sertorius,* whom Plutarch described as having heard about some Atlantic islands ca. A.D. 75, but he never actually went there. Neither Ducier nor the oracle is mentioned in critical studies of Sertorius.[11] The situation could spring from a now-lost historical

---

a gallo et portaa nile coste dil insula Tullia majore ove li marinari se detteroe hale femin haborigin / Sestorius et oracolu ritornoroe in Biscania . . . / Padua / 7 J' 1426 / T.V."

*The change to "Sestorius" might be attributable to Visco's having incorrectly copied Ducier, and indeed Visco at one place uses the spelling *Ducier* and at another *Dacier*.

trace separate from Plutarch or a fantasy of "what might have been" if Sertorius had actually sailed.

Page "C" shows the "after Deluge" version of the map, presumably as attributed to contemporary sailors. Visco's explanation on page "D" replaces classical pagan interests with concerns somewhat more contemporary:

> Duminicus Ducier, monk of Domremy, a scholar of ancient manuscripts of his own collection, designed a world map attaching the part broken off by the water of the deluge and earthquake. The said part of the globe broken off which found a place at a long distance in the ocean was previously referred to by him as Tullia Major island.
>
> Between this and Tullia Minor island there is a forest of seaweed in which birds live with a head like a man, hair, eyes, nose, mouth with two teeth, with wings like a scorpion and legs like a frog, but no body. This bird talks, sings, flies and swims. It goes into a delirium at the sight of mariners dressed in black.
>
> Ducier brought about, with respect to the much-acclaimed Joan of Arc, the merit of the serene King Charles VII, through the fact that he cursed the corpse of Bedford, and the people and clergy can see the ghost of Bedford suffer.[*]
>
> See the Chronicle of Besançon of the Holy Mother Church.
>
> 20 May, 1439.   Ts. Visco[†]

While Ducier's preoccupations do not impress us as being especially scientific, they were nevertheless par for his era. They are no less respectable than some activities of the highly regarded later scholar John Dee. In fact, the description, "a scholar of ancient manuscripts of his own collection" places him among the early humanists. The map which Visco has preserved for us is only a sketch, but it shows that there was a concern in his era about mainland in

---

[*]This presumably refers to the English archenemy John Plantagenet, Duke of Bedford, who was buried in the cathedral at Rouen in 1435. His contrivance in the execution of Joan of Arc created as much hatred as his sieges.

[†]A translation of: "Duminicus Dacier [*sic*] monaco di Domremy studioso di antiche scripture di soa collectione disegnoe un mappamundu coa parte rotta di le acqua dil deluviu et terremoti / dicto parte rotta dil globo trovoe logo a molto distanzo nil mare oceanus predicta da illu insula Tullia Major / et da dicta et insula Tullia minor havi foresta di erbe mare in habitata da uccelli co cocuzzo come homo / copeli / ohi / harechi / naso / ducca co bi denti / co ali come scorpion et gambe come rane / no corpu / dicto parla / canta et vala et nata / dicto va in deliriu nil vide marinai co nere vesti / Dacier foe di la molto honorata Jeanne D'Arco meritato dil sereni' C. L. VII / d' por questo facto illu maledissoe il corpu di Bedford et populo et clero vidi lo spiritu di Bedford suffrire / Vidi cronaca di Besancon di Sancta madre chiexia / dil 20 / majo / 1439 / Ts. Visco"

the west. If the manuscript is authentic, perhaps even the Sargasso Sea here finds its first cartographic representation, preceding Andrea Bianco's surmised depictions of 1436 and 1437.[12]

1424: North Atlantic Chart. This nautical chart (Figure 31) was signed by an otherwise unknown Zuane Pizzagano, the signature, although erased, being still discernible. The Venetian author's final purpose was to define sailing conditions in the open western ocean. He depicts the European seaboard and only the entrance to the Mediterranean, while devoting half of the map to the ocean itself.* Note that with the scale he has used the Grand Banks of Newfoundland would not be much beyond the end of his chart.

The two large islands shown in the center of the ocean, Antilia and Satanaxes, appear here for the first time. They were to become standard features of later maps and become heavily intertwined with various Atlantic island mythologies of St. Brandan, the Seven Cities, and others. However, their appearance here is not in any mythological context. The scholar Armando Cortesão suggested that they represent early Portuguese knowledge of the West Indies, which contain two primary islands, Cuba and Hispaniola. However, in attempting to account for various apparently nonsensical words inscribed on the islands he suggested the likelihood of Norse influence. The American historian James E. Kelley, Jr., thinks these are not islands at all but sections of the North American seaboard.[13] One of the smaller islands on this 1424 chart is given the name Ymana, certainly reminiscent of De Virga's Armana. The Irish legends have St. Brandan searching for an island named Yma, one of the Isles of the Blessed. Robert Fuson has given evidence supporting an interpretation of Antilia and Satanaxes as representations of Formosa and Japan brought via Venetian contacts with knowledge from the Far East.[14] If so, then this map demonstrates the growing tendency of cartographers to assume a global view that looked westward for the East.[15]

SUMMARY

It seems quite clear that southern Europe was now looking westward beyond the Atlantic horizon. Even if the early-fifteenth-century evidence by itself is not conclusive that the Norsemen explored or knew about the American eastern seaboard, that evidence begs to be taken seriously. The same information that constitutes that evidence, when the fifteenth-century geogra-

---

*Nevertheless, a significant portion of the skin to the east (rolled up in the photo) remains unused but has rhumb lines laid out on it.

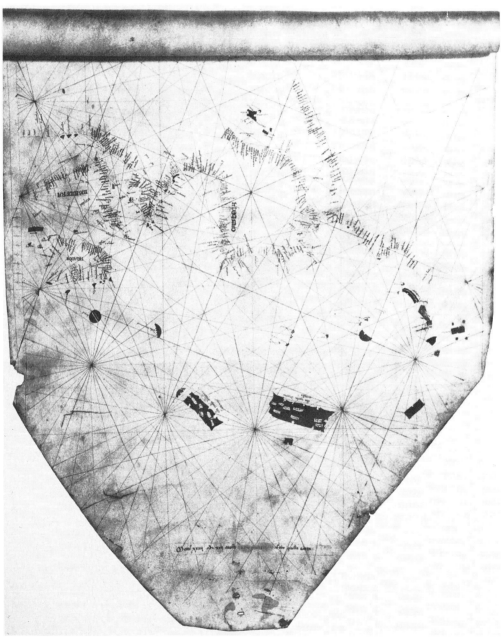

Figure 31. North Atlantic nautical chart by Zuane Pizzagano, 1424. Location: Minneapolis, James Ford Bell Library. Source: Armando Cortesão, *Nautical Chart of 1424* (Coimbra, 1954), frontispiece. Courtesy of Coimbra University.

phers took it seriously themselves, could have formed the foundation for the growing European interest in the westward approach to Asia.

We have identified maps with eastern seaboard origins stretching from the Quebec-Labrador peninsula to perhaps even the Caribbean islands and the Sargasso Sea and will identify more below. During the same period we saw an upswing of interest in global geography, exemplified by the Latin translation of Ptolemy and by Pierre d'Ailly's concerns. D'Ailly and Chrysoloras were clearly progressive thinkers in their time, willing at an early stage to take seriously traces of information that led to radical implications. More conservative thinkers simply rejected the new information at first. When, after a decade or so, some of them also started to deal seriously with it, they found interpretations with less radical implications, such as attributing the data to Scandinavia or the middle of the ocean. Without any explicit discussion, the divulgence-hiding paradigms appeared firmly established.

CHAPTER 11

NEWS PENETRATES THE ESTABLISHMENT

THE SCHOLARLY ESTABLISHMENT OF THE FIFTEENTH-CENTURY was probably just as inclined to ignore wild stories about Norse explorations as today's is to reject the controversial stories described in Chapter 9. Such stories (and similar ones of the same era not mentioned here) just cannot be made to fit into the accepted state of knowledge without changing it fundamentally. But gradually, in the form of rationalizations and misunderstandings, the fifteenth-century scholars apparently did proceed to assimilate the information.

1427, 1431: Claudius Clavus. Clavus made his first known map in 1427 (Figure 32) and a second in 1432, which, though now lost, is known through its surviving "Vienna text" and a nearly exact copy by Nicholaus Germanus (Figure 43). These were discussed in detail in Chapter 2.[1] There we saw that they seem to describe Norse contacts with information from Hudson Bay to Alaska. The first map is highly suggestive of Seward Peninsula and the Bering Strait, while the second seems to be based on Southampton Island as well as suggesting the Aleutian-Alaskan peninsula.

This Vienna text opens with what appears to be a signature of Clavus, his genealogy. This same genealogy was worked into the Nancy text accompanying the first map. However, while the Vienna text used this information as its introduction, the Nancy text found its excuse in pinpointing the geographical location of Clavus's birthplace in Denmark. At the corresponding point in the Vienna text we find: "In the middle of this island [modern Funen] is the village of Salingh, in which the author was born 14 September, 1388 A.D., at two hours before sunrise." Essentially nothing else is known of Claudius Clavus.[2] Not even the year or circumstances of his death are known. However, he had a lasting effect on cartography, and perhaps history.

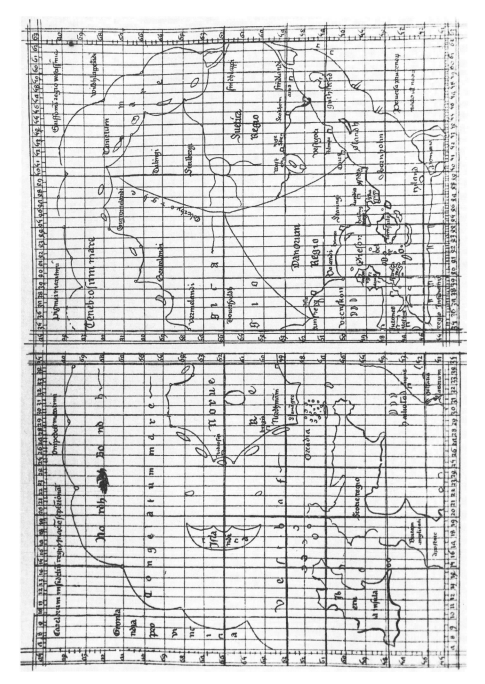

Figure 32. Claudius Clavus's 1427 map of the North. Location: In Filiastrus Ptolemy at Nancy, Municipal Library, Cod. Nanc. lat. 441. Source: Nordenskiöld, *Facsimile Atlas*, p. 49, fig. 27. Courtesy of Map Division, The New York Public Library, Astor, Lenox and Tilden Foundations.

Figure 33. Modern reconstruction from the Vienna-Klosterneuburg coordinate table "Cosmography of 7 Climates," 1425–50. (Northern archipelago corrected by author from coordinate table.) Location of coordinate table: Munich, Bayerische Staatsbibliothek, Cod. lat. Monacensis 14583, fol. 222r–235v. Adapted from Dana Durand, *The Vienna-Klosterneuburg Map Corpus* (Leiden, 1952): pl. 10, by permission of E. J. Brill.

Circa 1425–1450: Vienna-Klosterneuburg Map Corpus. This body of cartographic material was produced by several people working in and around Vienna from about 1425 to 1450. It was gathered into a single volume by a later collector. The corpus includes a few original sketch maps, but the most im-

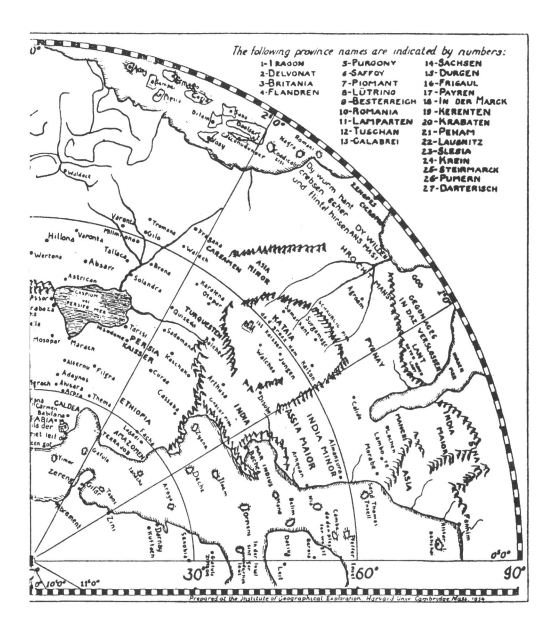

portant contents are extensive tables of coordinates from which maps could be reproduced after Ptolemy's suggestion. None of the finished maps are known to remain in existence, but in the 1950s Dana Durand reconstructed them from the coordinate tables. (Incidentally, the type of coordinates used is not global latitude and longitude. It is a flat polar coordinate system.)[3]

The two maps shown in Figures 33 and 34 are based on Durand's reconstructions. The first map, derived from a table called Cosmographia Septem Climatum, shows an archipelago nestled along the northern coast of Asia.

NEWS PENETRATES THE ESTABLISHMENT   167

Figure 34. Modern reconstruction from the Vienna-Klosterneuburg coordinate table "New Cosmography," 1425–50. Location of coordinate table: Munich, Bayerische Staatsbibliotek, Cod. lat. Monacensis 14583, fol. 236r–277v. Source: Durand, *Vienna-Klosterneuburg Map Corpus,* pl. 13. Reprinted by permission of E. J. Brill.

This map has been slightly modified from Durand's reconstruction. For some unstated reason Durand fused these islands into a single large island in his own reconstruction. However, the coordinate table clearly denotes by marginal symbols that they are part of a whole list of individual islands. This is

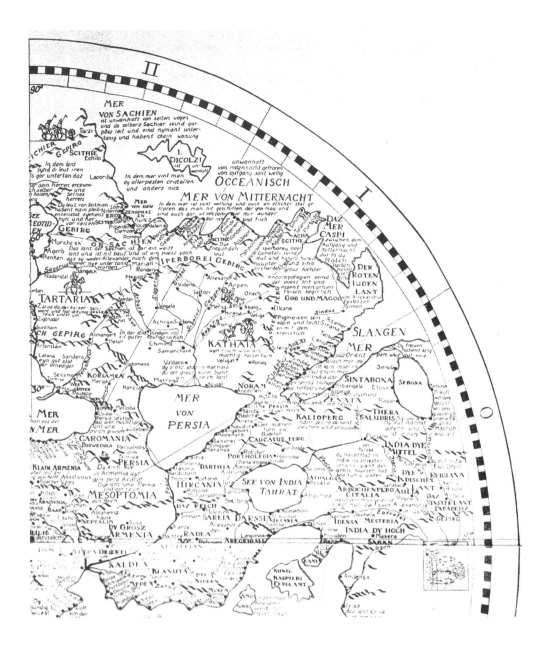

evidently the Arctic Archipelago, now firmly transferred from the western ocean to the north of the world continent. The names sound almost European (Romani, Nagra, Coadicolis zisi, Daosler, Caenconudinper, Saba, Vosy, Dilen, Coris, Ermago, Pirsia, Acoy), but they still have a foreign or artificial aura. One is also reminded of Salmos and Naron, the Arctic islands with southerly names in the Catalan fragment at Istanbul (page 140). Perhaps they all have a *raison d'etre* similar to Clavus's artificial names (page 48).

The second map, based on the table referred to as Nova Cosmographia, the New Cosmography, has the archipelago give way to a single large island, Dicolzi. The island Dicolzi bears a general resemblance to both the Vinilanda of the Yale map and the later Grocland, namely, they are all long and narrow, whereas the typical invented island is often formalized as circular, trefoil, or squarish. I have pointed out elsewhere that the name Dicolzi might further corroborate the pastureland interpretation of the meaning of Vinland.[4]

Also apparent in the maps is a Sea of Red Jews in eastern Asia and its associated Land of Red Jews. The legend of the Red Jews is closely associated with the legendary lands entered by descendants of Noah after the Flood. These lands turned out to be inhabited by hordes of warlike barbarians who installed Noah's grandsons Gomer (Gog) and Magog as their leaders. When Ezekiel prophesied the reuniting of the scattered Jews in Israel, he also had a prophecy for the lands of Gog and Magog. It was that they would rise up in a great battle of Armageddon. A preoccupation of geographers down through the ages has been to pinpoint the unknown location of Gog and Magog with hopes of averting this catastrophe. Alexander the Great is reputed to have built a vast iron gate across a mountain pass through which, it was feared, the warriors might come.

Geographers located Gog and Magog predominantly in the Middle East or far north until early in the fourteenth century. Then, after the Mongol invasions, mapmakers showed them more predominantly in the Far East. There is a remote possibility that the divulgence-hiding paradigms were also involved in this shift. We may look for a hint in the effect that Norse contact with the Eskimos apparently had on reviving the myth of the Arimphians and Hyperboreans. Any Norse contact with eastern seaboard Indians could not fit that myth, but it could fit the Gog-Magog stories of warlike tribes. In this interpretation, the displacement of Gog and Magog to the Far East would include an educated interpretation of Norse travelers' beliefs that they had reached the shores of Asia. Almost simultaneously with this shift arose new, previously unheard of references in Germanic literature to a lost tribe of "Red Jews." The first appeared in 1280, associated with Gog and Magog. There are historical linguistic aspects to this phenomenon, but a reality-based component is also possible.[5] The Beothuk Indians were the most accessible to Norse contact, and they are famous for their use of red paint and are the source of the concept "Red Indians." We shall see further development of these ideas.

1436: Andrea Bianco. A highly renowned Venetian shipmaster and mapmaker, Andrea Bianco published in 1436 an atlas of modern sailing charts. To this he appended his world map in the old circular style (Figure 35) as well as a copy of Ptolemy's world map. This juxtaposition of portolan theory, wheel map tradition, and Ptolemaic theory signals that Bianco was an acute thinker about global geography and interested in stimulating the thinking of his contemporaries.

The circular world map shows an island in the same place as the Dicolzi of the Vienna-Klosterneuburg corpus, and on it is the notation "griffons and girfalcons." However, this map's treatment of Gog and Magog is different from that on any other map. Bianco depicts this land as an extension of Asia which juts out into the blue border surrounding his map, as if beyond the middle of d'Ailly's equinoctal circle. Part of Bianco's representation of Taprobana (Ceylon) also extends into the other hemisphere by use of d'Ailly's device. However, it does not appear that Bianco clearly understood or intended this map as a global representation in the same sense as d'Ailly did his Seventh Figure. While Bianco's meridian line passes through the same area as d'Ailly's, his equatorial line passes through Greece, Italy, and Spain, far north of the true global equator. The curious mixture of wheel-map tradition with global projection theory is further emphasized in the west. There Bianco apparently gives explicit attention to d'Ailly's ocean problem by showing the full extent of the ocean to the global horizon. This is a device that appears again below and seems to be part of a transitional phase in which consideration was being given to the situation of the world continent relative to the global ocean.[6]

Circa 1440: Vinland Map at Yale University. When this Vinland map (Figure 36) was first discovered, its authenticity was challenged on various grounds. One was that the delineation of Greenland was too accurate to have been possible in 1440, particularly the fact of its insularity. But we see in his Vienna text that Claudius Clavus was aware of that insularity. Presented below is further evidence of contemporary awareness of Greenland's insularity. The Eskimos, who had migrated down each coast from the northern extremity, knew of it. In fact, the Yale Vinland Map fits its asserted position in this cartographic chronology so neatly that its accuracy now confirms rather than undermines its authenticity. It also shows a Norse ability to make or inspire maps; the nearly perfect outline of Iceland that it includes could only have been surveyed by or with a Norseman, as Eskimos never went there unless escorted by Norsemen.

The most striking inscription on the map is in the upper left corner:

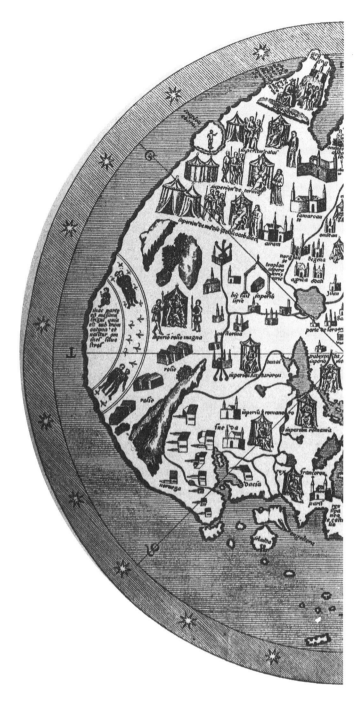

Figure 35. World map by Andrea Bianco to accompany sailing charts, 1436. Location: Berlin, Deutsche Staatsbibliothek, Cod. Hamilton 108. Source: Tracing by Fornaleone

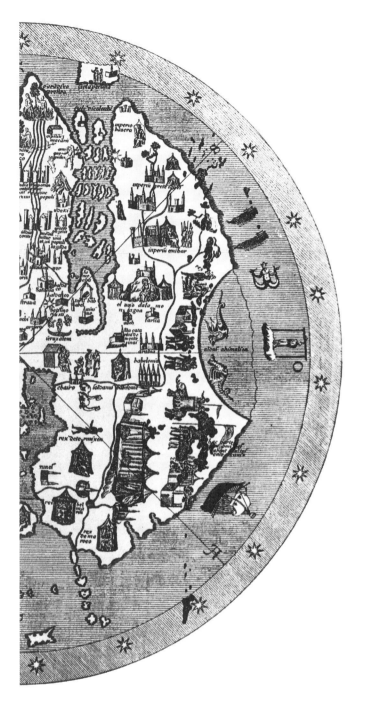

in Nordenskiöld *Periplus,* p. 19, fig. 7. Courtesy of Map Division, The New York Public Library, Astor, Lenox and Tilden Foundations.

Figure 36. The Vinland Map at Yale University, ca. 1440. Location: New Haven, Bei-
necke Library. Source: Skelton et al., *Vinland Map,* following p. 16. Reprinted by permission of Yale University Press.

By God's will, after a long voyage from the island of Greenland to the south toward the farthest remaining parts of the western ocean sea, sailing to the south
amid ice, Bjarni and Leif Eriksson as companions discovered a new land, most fertile and even bearing vines, which island they named Vinland. Henricus, the legate
of the Apostolic See and bishop of Greenland and the neighboring regions, arrived in this really spacious and most opulent land in the last year of our most Holy
Father Paschal [1117–18] in the name of God Almighty. He remained a long time

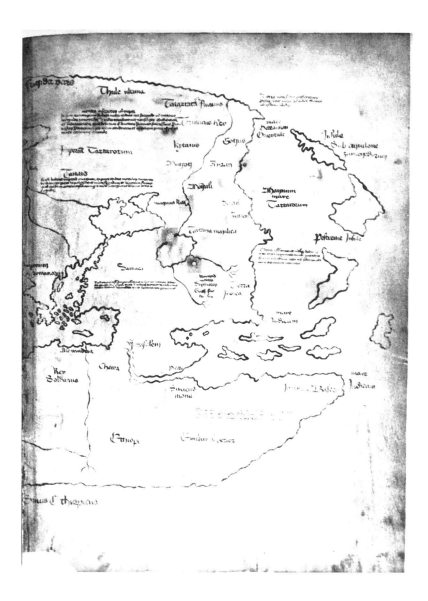

summer and winter. Later he returned toward Greenland and then in the most humble obedience to the will of his superiors proceeded southeastward.

As I have shown elsewhere,[7] this inscription incorporates a double acrostic naming Bishop Erick Gnupsson and his faith.

The inscription's text substantiates information in existing copies of older sources and adds a few new details, such as the length of Gnupsson's stay (page 91). However, there is a fine point of wording to which we should pay closer attention: The phrase "toward the farthest remaining parts of the western ocean sea" is curious. Without knowledge of a continental land mass to the west, wording like "toward the farthest remaining parts of the *earth's lands*" would have been expected rather than "ocean sea." (Either could have been expressed without affecting the covered acrostic.)

The cartographer's awareness that Greenland is *not* connected by a land arc to the Eurasian continent allows him to consider mainland to its west. His understanding that Vinland lay toward the farthest *remaining* part of the ocean suggests that one should view the map as continuing from the left edge immediately to the right edge, that is, we can presume that the map depicts the entire globe, including the ocean, going a step beyond Pizzagano's 1424 North Atlantic nautical chart. Wile the main body of this map was copied from Bianco, the easternmost extreme is not. The coastline of the East as depicted here has no known prototype, and it could have served as a prototype for some later maps, such as the Genoese map of 1457 (examined below) and those of some of the later Ptolemies. The most unusual feature of this farthest remaining part of the ocean sea is, of course, the three large islands which replace the strange promontory on Bianco's map.* The cartographer gives them a most unusual name: the Postreme Islands. This Latin word means almost the same as *extreme* but carries with it a connotation of "behind" or "rearward." The cartographer seems to have been thinking of these islands as being around the back of the globe, "down under." Another inscription, above these islands, identifies them as isles of the Samoyedes *sub aquilone,* that is, "at the North Pole." This opens up a possibility for a surprising interpretation of these islands, not consistent with their apparent latitudes. It is possible that these represent the Arctic Archipelago, reduced to three main islands, as in de Virga. The cartographer has placed them at the postremes of the earth, but the latitudes of the inscriptions are still on a parallel with Greenland and Baffin Island. He has evidently identified the Eskimos as the Samoyedes of Siberia mentioned in the Tartar Relation accompanying the map. At the very top of this area the cartographer left the following comment: "Lands not sufficiently thoroughly explored [*perscrutate*]. They are situated among the northern ice which isolates them." The compiler of this map evidently not only thought that the Norsemen had cir-

---

*While Japan had been heard of through Marco Polo, no maps of it reached Europe (unless it is the island named Satanaze in the 1424 North Atlantic chart, as Fuson suggests). Aside from Greenland and Vinland, these three islands are the most prominent ones on the map. Some writers believe they are merely fortuitous results of erroneous copying (Wilcomb E. Washburn, "Representation of Unknown Lands in Fourteenth-, Fifteenth- and Sixteenth-Century Cartography," p. 15, *Agrupamento de Estudos de Cartografia Antiga,* vol. 35, Coimbra, 1969).

cumnavigated the Ocean Sea but he also was aware that they were systematically exploring their finds.[8]

1448: La Sale's **Salade.** When the French king of Naples, René d'Anjou, needed a tutor for his son, he found the perfect blend of diplomatic experience, soldiering, and academic knowledge in his countryman Antoine de la Sale, a knight. La Sale recorded some of his wide-ranging teachings in a book he called *La Salade* in 1442. In it, in the midst of a tightly summarized survey of world geography, he lapsed into a protracted discussion of the Greenland area. Some of the material, especially regarding animals, may seem fantastic, but most of it can be interpreted as based on real evidence.

There is an island called Iceland, where there are the countries named Greenland and Unimarch, with large numbers of bears that are all white. Item; in the Norwegian sea, opposite to our midnight, there are diverse islands erupting fire and stinking flames; from these come the sounds of many crying, bawling voices, moans of lamentations as well as terrible voices, thunder and lightning; just as St. Brendan wrote, and of one island that carries monks who lead saintly lives; and in another isle, named People of the Alb, a great abundance of people like spirits who are incessantly purifying themselves there, making processions in honor of God continuously through the entire day and night. And of this part of Norway we tell further of the great marvels that are here; that extend toward the ends of the Occident to the Russians and to the Tartars; and in the domain of the pole, where the north wind blows to the sea from the land of ice, so-named because in winter time ice extends from these lands forty leagues (of the type with forty cubits) into the sea; which lands do not become extricated again until only in the month of June, on account of its heat and the impetuous winds that it has blowing and the great power of its weather. The winds blowing are now called *Estand,* and formerly they were called by the poets, *Chilte.*

And here is the largest island in the world, in which the people are very sharp and industrious, strong and robust of limb and adept with weapons and of a religious inclination. And so there one finds a very great and copious multitude of cattle, grassy fields and highly fertile pastures. Wheat never grows there, nor any other grains suitable for human usage. They catch infinite numbers of fish, which they dry in the sun; and of this they eat and make their bread. They have no knowledge of wine or cider; water and milk are their beverages; all fruits are unknown to them. There are mountains there of such very high elevation that they reach almost to the middle region of the air. And from there descend rivers that are of such a virtue that all the boughs and wood that have touched their waters, the part that they contact is set on fire, while the rest remains unaffected. In crevasses of these mountains originates the glacier that has remained frozen solid there ever since ancient times. There are various kinds of suffering there, for which there are gushing springs of hot water, most suitable in several diverse maladies; and there is mining for silver, copper and lead; and in certain played-out mines live birds with high crests and tails in the manner of roosters, which have a chamber in their belly enclosing large clear pearls; but few are found to be well-rounded. In a couple of these mountains there appear great rocky areas where a single plant does not grow; and there live falcons white as swans and rather more aspiring than those of ours. There are no wild animals there except white bears large as oxen and spotlessly white foxes, that at times are carried on the ice as far as to Iceland. One sees monstrous fishes there that are very marvelous; for, in the foreparts some have the appearance of horses, others of cows, stags, goats, dogs, others resembling men and women to the waist besides, and from the waist onwards are fish scales. There are whales there a hundred cubits in length, that have four great teeth above and four below, which are two cubits long and a thickness to match. The ponds of Greenland give birth to lampreys that descend as far as the utmost end of the sea, that are thirty cubits long, or more sometimes, and have a cubit of circumference as well as around the head, and often they attach themselves one to another lengthwise and then bind up a vessel in the sea, that, if it is not quickly shaken loose, will soon be lost. There are dolphins there, of the same type as elsewhere, except that they are somewhat larger and bear such love and affection for humanity that if they see anybody in the sea they come together underneath him and, whether he be living or dead, carry him to the nearest land; and for this reason they are protected, by very heavy penalties, that no one eat dolphin or do them harm, on pain of sacrilege, and likewise that no aid and assistance be denied them, if required, from extreme conditions of the sea or otherwise.

Some of the by-gone astrologers placed part of this region beyond the climates, on account of the very bitter and protracted coldness there, overpowering the beams and warmth of the sun; as a result of this the land there is thus rather barren and unprofitable. Those realms and parts [subject to] Norway terminate at the great sea towards the south; and in those parts live savage and monstrous beasts too horrible to put in writing; which are towards the market towns of *Holcode, Endeust* and *Sindoren,* which face towards the countries of Great Britain and Scotland and towards the land of France, where Hercules placed his columns; and there there are great islands of cultivatable fields and abundances of things to eat.[9]

The statement that Greenland and Unimarch are part of Iceland seems to be a variation on the provenance paradigm, placing unfamiliar lands as part of Iceland instead of Scandinavia. The unfamiliar name Unimarch is without meaning in the Norse language. If the word is taken as Antoine de la Sale's French, it can be interpreted literally as "flat stepping stone," probably referring to the Helluland of the sagas.* This is usually taken as Baffin Island. La Sale's Unimarch would correspond to Cnoyen's Grocland and the Vinland Map's Vinilanda, all paired with Greenland, further corroborating all these identifications.

The reference to volcanic islands in the area of Iceland is likely based upon fact. If the islands were inhabited when the eruptions occurred, supernatural interpretations of the anguished voices are not needed. The island named People of the Alb (a white vestment) likely refers to the Norse *Hvitramannaland* (White Ones' land), an Irish outpost where the records described people going about in great processions wearing white robes.

The true characterization of Greenland as "the largest island in the world" incidentally establishes that southern Europeans of this era had the information that Greenland indeed was an island. Such information was available to the Norsemen from the Thule Eskimos, who had migrated around both sides of Greenland from its north shore.

The rivers that set wood afire have no reality in Greenland, but lava flows in Iceland do behave according to the description given. The bestial descriptions need not detract from the trustworthiness of the account, as they were standard fare of the era. Nearly a century later the highly regarded map of Olaus Magnus (Figure 66) brought these monsters to life pictorially.

The transatlantic market towns *Holcode, Endeust,* and *Sindoren* seem to be named with neither Norse nor French words. Remember that de Virga also named three places there, one of which had the same name as Marco Polo's market town *Singui.* Claudius Clavus also spoke of a market town, his *Ynesegh.* The naming by three different sources of market towns facing across the ocean toward middle European latitudes suggests that the Norsemen might have been active on the American continent. The world view here is also globally trans-Atlantic, rather than just keeping to the home circle.

---

*I have not been able to find *Unimarch* in any dictionaries, but *marche* means "step" or "tread," and *uni* means "smooth, even, or level." In the context, equivalence with the Old Norse *hellusteinn* and thus Helluland is quite plausible.

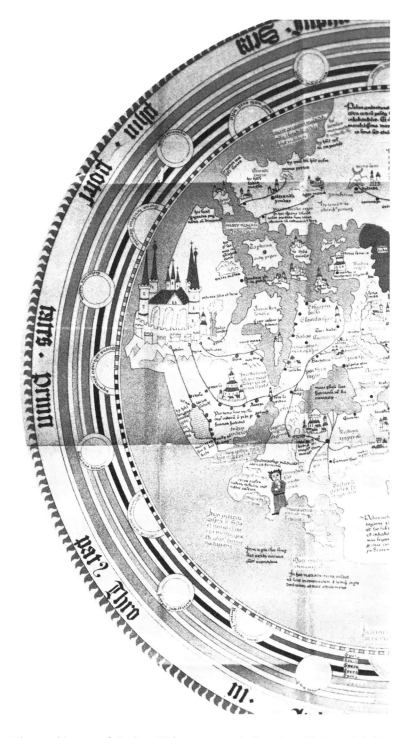

Figure 37. The world map of Andrea Walsperger, 1448. Location: Vatican, Pal. lat. 1362B. Source: Konrad Kretschmer in *Zeitschrift des Gesellschaft für Erdkunde zu*

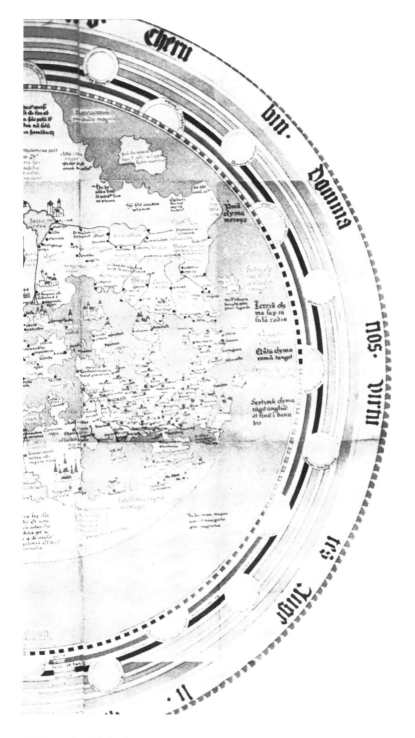

Berlin, vol. 26 (1891).

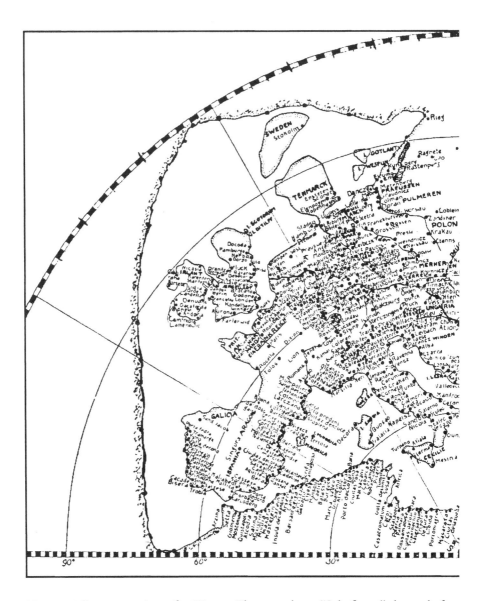

Figure 38. Reconstruction of a Vienna-Klosterneuburg "Schyſkarte" drawn before 1450. Location of coordinate table: Munich, Beyerische Staatsbibliothek, cod. lat. Monacensis 14583, fol. 300r–312v. Adapted from Durand, *Vienna-Klosterneuburg Map Corpus,* pl. 17, by permission of E. J. Brill.

1448: Andrea Walsperger. Unknown aside from this map (Figure 37), Andrea Walsperger worked in Salzburg. The Walsperger map is closely derived from the Vienna-Klosterneuburg school but the orientation of Walsperger's map is rotated 180° from that of the New Cosmography. It shows in a contemporary hand the island Dicolzi, which is completely alone, with no neigh-

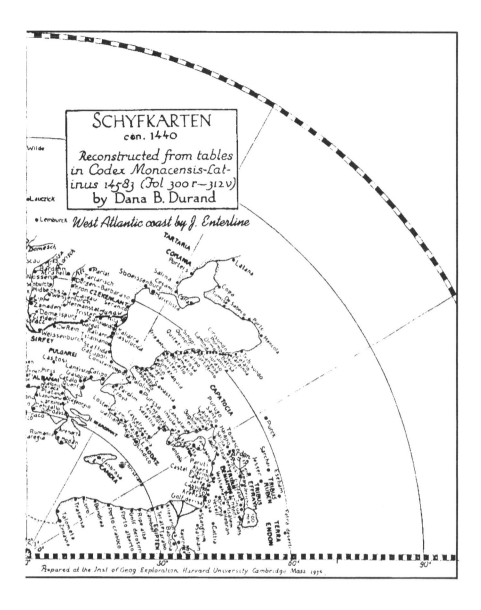

The map contains the following inscription in a boxed title:

SCHYFKARTEN
cån. 1440

Reconstructed from tables
in Codex Monacensis-Lat-
inus 14583 (Fol 300 r—312 v)
by Dana B. Durand

West Atlantic coast by J. Enterline

Prepared at the Inst of Geog Exploration Harvard University Cambridge Mass 1936

boring islands shown. Note for later reference that the disk of the Earth's land is displaced off-center in the circle of the surrounding ocean, but in a different direction than on Bianco's map.[10]

Before 1450: Vienna-Klosterneuburg Schyfkarte. This reconstructed sailing chart (Figure 38) is described in the later tables of the Vienna-Klosterneuburg map corpus.[11] The most arresting feature of the map is, of course, the unbroken coastline running from Lapland out into the Atlantic and southward to connect with Africa. The coordinate table from which the map was con-

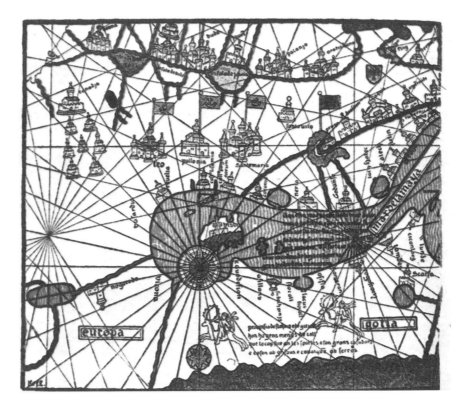

Figure 39. Northern portion of a Catalan map in Florence, 1440–50. Location: Florence, Biblioteca Nazionale Palatina, Port. 16. Source: Nansen, *In Northern Mists*, vol. 2, pp. 232–33 (tracing by Björnbo).

structed treats this coastline very matter-of-factly and does not hesitate at the idea of mainland in the western ocean. Durand, however, rendered it with tentative dashes; the uniform cartographic rendition of this feature from the tables is supplied by me. One is immediately reminded of the idea expressed in the *Icelandic Geography* of 1159 and in the *Gripla* that Vinland is connected to Africa. However, the Vienna-Klosterneuburg school in general and particularly in their Schyfkarten showed an aversion to reliance on mythical or legendary sources. They generally demanded concrete data for their representations. At any rate, this chart suggests that awareness of the old Norse stories was not restricted to the North. Perhaps the hard data on the land mass adjacent to Greenland was sufficient to lend credibility to the old legends but not enough from which to plot a detailed coastline. The conceptual framework suggested here is similar to that in Ducier's pre-Deluge cosmography (Figure 30), that of an inland-Atlantic rather than of the one-

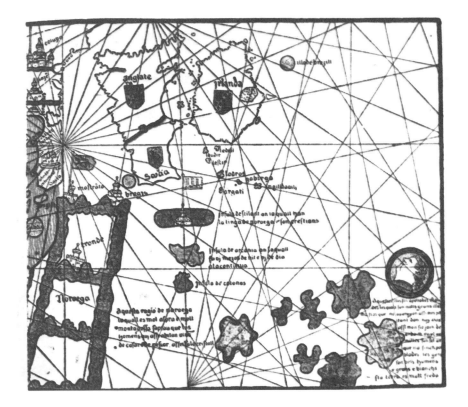

ocean paradigm. This could have been a separately developing new paradigm in the fifteenth century, forerunning Björnbo's idea.

1440–1450: Catalan Map in Florence. This anonymous compass chart (Figure 39) with inscriptions in Catalan is kept in Florence. It is apparently based on the Dalorto map of 1325 (Figure 23) and previous Catalan maps but was updated with modern information. While the Fortunate Lake, the Isle of Brasil, and Dalorto's Insula Tille appear at the same places, Scandinavia's extension to the west has been reduced and instead a large group of islands has been introduced. The partly illegible inscription* says that they are called "Icelands" and describes them as very large and very cold. It says that five of them are inhabited by attractive people, large and white. This could refer to bleached-hide Eskimo costumes, always bulky but often attractively deco-

---

*Partly with the help of the Institut Cartogràfic de Catalunya, I have attempted transcribing the inscription thus: "Aquestas illes son apellades islan / des tes quals son molts grans illes / ma ssos que no aparagon assc mes per / tant don noy ???? / assiman sio jaen de / ardoim vuyt ?? / milles son set an / que no sinch poi / blades. les gens / son bels homens / e grans e blanchs / —sta terra es molt freda"

rated. This seems rather strongly to point to information on the Arctic Archipelago, placed here as part of the Scandinavian sphere.[12]

Circa 1450: Catalan Mappemonde in Modena. A wheel-shaped chart (Figure 40a) kept in the Royal Este Library in Modena* seems to be based on the same sources as the Catalan map in Florence.[13] This map, like the Walsperger (Figure 37) and Bianco (Figure 35) maps, has the disk of the Earth's land displaced off-center in the disk of the sea. Thus, at one side the lands touch or are tangent to the border while at the opposite side the extent of the sea is emphasized. In earlier times the Orbis Terrarum was believed to be a perfect domelike cap emerging from a sphere of water. An assumed miraculous off-centering of a sphere of solid land within an independent shell of water was a means of explaining why the sphere of land was not completely inundated by the supposedly spherical shell of water.

This off-centering reminds one of the statement in the inscription in the Yale Vinland Map that Vinland lay "toward the most distant *remaining* part of the ocean." The Walsperger map showed no evidence of the Arctic Archipelago or of Norse activities in the western ocean, but rather in the north. Its displacement is accordingly such as to make the ocean widest in the north. The Catalan map in Modena, on the other hand, does show the archipelago in the west, and Bianco shows his strange promontory in the east, while the Vinland Map shows its archipelago in the east. The latter three accordingly make the western ocean the widest. This correlation again suggests trans-global, one-ocean thinking by the cartographers.

The archipelago (Figure 40b) is shown on this map in more detail than on the map in Florence, and the individual "Icelands" are given names such as Donbert, Tranes, Tales, Brons, Mmau, Bilanj, etc. This Catalan map (Figure 40a) also shows inherited evidence of the archipelago in the north and has an Arctic bay with the same two islands, Naron and Salmos, as the Catalan fragment in Istanbul (page 140). In this map, however, Naron though not easily discernible at this scale of reproduction, has the unusual outline suggestive of Victoria Island, showing the indentation of Prince Albert Sound (see front map). The group of three islands farther east bears an inscription to the effect that "In these waters live very beautiful girfalcons" and attributes the land to the Grand Khan of Marco Polo's description. They are clearly de-

---

*Some library catalogues have confused this with a 1375 Catalan *atlas* held in the same Modena library.

rived from the same two falcon islands as on the 1380 fragment, which I previously surmised as representing Greenland and Baffin Island.

1456: Gunnbjörn's Skerries Disappear. When Erik the Red discovered Greenland he was investigating some low-lying, tidally washed islands west of Iceland. These had been discovered a generation before by another Icelander, Gunnbjörn. Widely known as Gunnbjörn's Skerries, they were to become a landmark on the standard voyage from Iceland to Greenland. They were supposed to be situated just midway between the two. They also served as hideouts for outlaws throughout Norse history.

As today's map shows, there simply are no such islands between Iceland and Greenland. The key to this situation seems to be provided by the 1507–8 map of Johann Ruysch (Figure 54), which shows an island in the proper place with the inscription "This island has been totally exploded in the year 1456." The volcanic activity for which the Iceland area is famous is usually known for building new islands, but the reverse is also quite possible. In 1968 an entire island in Indonesia disappeared during volcanic disturbances, and many other examples are know. Subsequent to Ruysch, maps as late as 1700 showed shoal spots between Iceland and Greenland with the name Gombar Scheer. The feature has now subsided (via magma withdrawal) to a mere reduction in ocean depth, a rise from surrounding depths of hundreds of fathoms to within sixty fathoms of the surface.[14]

1457: Genoese (Florentine?) World Map in Florence. A map by an unknown Genoese (some writers say Florentine) author (Figure 41) carries in the western Atlantic a red title scroll, "This is a true description of the world of the cosmographers, accommodated to the marine chart, from which frivolous tales have been removed. 1457."[15] The unusual lenticular form[16] is necessary to accommodate the unprecedented extension of Asia to the eastward, reminding one of the oval form of the Yale Vinland Map (Figure 36). In fact, the easternmost coastline of Asia is, in its gross features, reminiscent of that on the Yale Vinland Map. Neither of these maps has any known prototype for the shape of that coast, not visited by Europeans until much later. Old-style geographical disk theory would have predicted this coast to be convex rather than flat or relatively straight. The cartographer provides two mileage scales just outside the border, one in units of 50 and the other of 100 miliaria. By these scales the extent from the leftmost apex of the lens to the rightmost is some 12,000 miliaria. That is exactly half the approximately cor-

a

Figure 40. Catalan map in Modena, ca. 1450. Location: Modena, Biblioteca Estense C.G.A. 1. (*a*) Wheel-shaped mappamundi. Source: Gasparini-Leporace, *Mostra "L'Asia nella Cartografia degli Occidentali"* (Venice, 1954), pl. 6. Reprinted by permission of the Italian Ministry for Cultural Heritage and Activities. (*b*) Northwest detail. Source: Skelton et al., *Vinland Map,* pl. 11.

rect fifteenth-century estimate of the circumference of the globe. The drawer of this half-world map was making allowance for a greatly extended eastern Asia but simultaneously denying d'Ailly's hypothesis that Asia extended around the back side of the globe. Particularly, he was denying the possibility of a "narrow" ocean between Spain and Asia.

This cartographer's eastward extension of Asia was based on actual travel reports about the southern edge of Asia brought back by Niccolò de' Conti in 1444. The extension left him with a gap to fill in the otherwise unknown

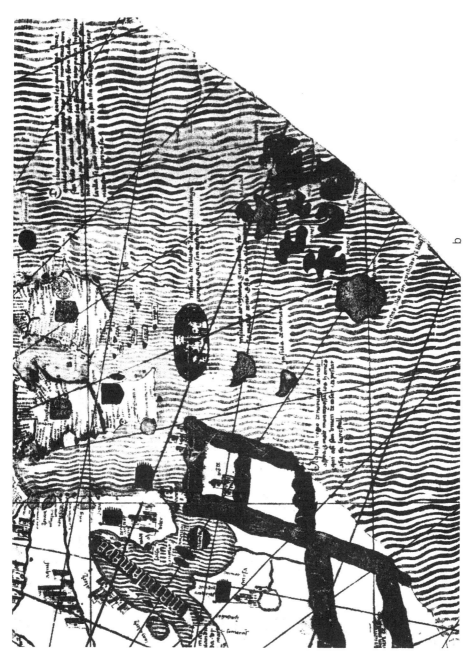

Figure 40. (*Continued*)

b

NEWS PENETRATES THE ESTABLISHMENT   189

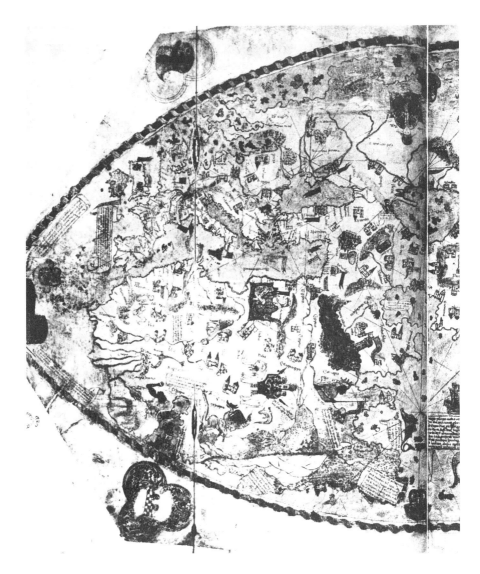

Figure 41. World map with inscriptions in Genoese, in Florence, 1457. Location: Florence, Biblioteca Nazionale Palatina, Port. 1. Source: Kamal, *Monumenta Cartographica,* vol. 5, fasc. 1, fol. 1494.

northern part. In this area we see a wedge outlined by two stylized mountain ranges which meet the coastline at pronounced inlets. These stylized mountains probably indicate the borders of the legendary enclosed nations of Gog and Magog. However, the cartographer's approach has been quite scientific otherwise, and we should not think he simply made up the shape. We might look for a prototype from Norse contact. The wedge-shaped Quebec-Labrador peninsula that appeared in de Virga's work seems to fit the circumstances. Note that the entirety of this wedge on the map is covered

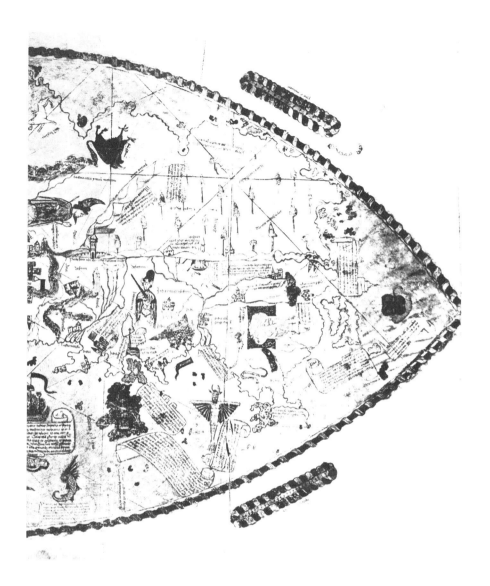

with trees. It is true that Siberia contains a large forest, but so do other places in the world. These are the only trees on the map. Recall that the literal meaning of the Old Norse name Markland, assumed by many to be Labrador, is "land of trees."

Off the coast of this wedge is an archipelago. At its northern extremity is a picture and inscription of a Hyperborean. Just as the Eskimos of the Arctic Archipelago apparently were interpreted to be the Hyperboreans, so perhaps were the native Americans of the Quebec peninsula interpreted to be the barbarians of Gog and Magog. This would seem the earliest clear example of subarctic map displacement under the one-ocean paradigm instead of the provenance paradigm.

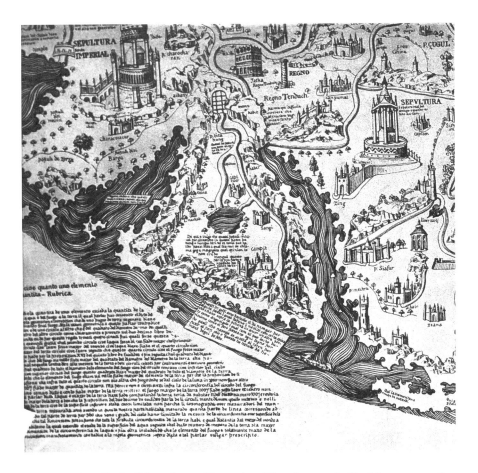

Figure 42. Detail of a disk-style world map by a Venetian monk, Fra Mauro, 1459. Location: Venice, Biblioteca Nazionale Marciana. Source: Santarem, *Atlas,* plates 4x–4y. Courtesy of Map Division, The New York Public Library, Astor, Lenox and Tilden Foundations.

Just below the wedge is a gulf, previously mentioned, and just below that is a bulbous prominence similar to that on Bianco's map (Figure 35). This prominence occurs persistently on later maps, eventually in a refined form that can be identified.

1459: Fra Mauro. There is a map nearly six feet in diameter, the northern portion of which is shown in Figure 42, that is considered to be the final work of a Venetian monk named Fra Mauro. It was completed with the help of co-worker Andrea Bianco in 1459. This is the last map of any importance still in the old disk school following the revival of Ptolemaic theory. Evidently the decision to retain the disk theory came to Fra Mauro because he had difficulty integrating the Norse lands in the west into the Ptolemaic

scheme, with which he was sufficiently familiar. In an inscription just off Scandinavia he says: "I follow such a cosmography as this because if, in the style of Marinus [Ptolemy's predecessor], one wanted to observe his meridians or parallels or degrees, it were necessary to mention how much, in respect to the explanation of the known part of this circumference, it would depart from many provinces of the original old Ptolemy; throughout all the highest zodiacal extent between the East and the North he calls *terra incognita* just because at his time it was not known to him." Fra Mauro seems to feel that he himself is sufficiently familiar with that north Asian coastline to take issue. Just west of the North Pole he again launches into a long apologia for not following Ptolemy, but he ends with, "To my credit I must say that I have tried to verify my findings through many years of personal experience and the experience of others who have seen personally and who have related to me these facts."[17] Facts? First, just below the North Pole, he shows a landlocked sea with the following inscription, "Here at the time of winter the ocean itself freezes for about 1,000 miles." There is only one inland body of water in the world that fits this description, Hudson Bay. It stretches exactly 1,000 modern English miles from Baffin's Foxe peninsula into James Bay, along the edge of the Quebec peninsula.

Just beyond the pole, Fra Mauro shows a two-horned, or hammerhead, peninsula bearing the inscription, "Here one finds the best girfalcons." Recall that we have just seen a good representation of Victoria Island in the cartographic record (page 186), which has been suggested as Behaim's Island of White Falcons. This hammerhead peninsula might be interpreted as an earlier representation than Behaim's of Kent Peninsula abutting Victoria (see page 59).

The next southerly peninsula is a larger wedge-shaped one, shown as Gog-Magog. A temptation to identify this with the Quebec peninsula as shown on earlier maps is encouraged by the fully developed waterways along its flanks. The Quebec peninsula is indeed the ultimate northeast peninsula of North America. Off its shore is an inscription warning of off-shore sea life that attacks the wood of ships' hulls, a problem mentioned in the Vinland stories.

SUMMARY

The maps of the mid-fifteenth century (including those examined in Part I) admit an interpretation that is becoming difficult to dismiss lightly: that contemporary Norsemen and/or their guests were exploring and/or inquiring intensively on the western shores of the Atlantic. Several contemporary European scholars seemed almost consciously aware of the fact. For the first time in a century and a half, scholars developed a strong enough interest in things specifically Greenlandic to call forth into the historical record items such as the Yale Vinland Map and La Sale's *Salade*.

CHAPTER 12

EUROPE'S WESTWARD AWAKENING

IF THE MID-FIFTEENTH CENTURY WAS THE PERIOD when scholars had to accommodate to news of the Norse contacts, then the latter part of that century was the time for practical men to act. The period is filled with plans and voyages. It also marks the beginning of an era from which many other documents as well as maps have been preserved. Some of them offer less excitement than do the graphical insights from maps, but they will reward with other kinds of insights. Particularly, they tie known Bristol and Portuguese explorations to the Norse activities.

1472–1476: Scolvus, Pining, and Pothurst. As the Renaissance matured, so did political relations in Europe. Denmark had some time since become the ruler of all Scandinavia. That included Iceland and, if only on paper, Greenland, and whatever lay beyond. She had also entered into an alliance with Portugal. That country, under urging by Henry the Navigator, had searched southward along Africa for a route to the oriental paradise. Portugal gave recognition to Denmark by including Scandinavian navigators and commanders on some of these voyages.

In 1472 King Christian of Denmark was persuaded, according to one interpretation by the Portuguese king, Alfonso V, to explore the neighborhood beyond Greenland. At a minimum, this implies awareness that there *was* a neighborhood beyond Greenland. The sketchy traces of the expedition suggest that it contained several ships, including those of the Norwegian nobleman Didrik Pining and his associate Hans Pothurst. It may have been under the command of the Norwegian pilot Johannes Scolvus. The Portuguese sent along as observer a nobleman, João Vaz Corte Real.

Very little is preserved of the activities or results of any such expedition.

However, the expedition evidently lasted long enough for a later map[*] to attribute Scolvus's voyage to 1476. Sebastian Cabot, referring to the three sons of João Vaz Corte Real, who later followed up on their father's presumed experience by searching for a northwest passage, preserved some illuminating information: "But to find oute the passage oute of the North Sea into the South we must sayle to the 60 degree, that is, from 66 unto 68. And this passage is called the Narowe Sea or Streicte of the Three Brethern; in which passage, at no tyme in the yere, is ise wonte to be found. The cause is the swifte ronnyng downe of sea into sea. In the north side of this passage John Scolus, a pilot of Denmerke, was in anno 1476." At that time, of course, Norway was part of Denmark. The strait described conforms well to Hudson Strait (see front map and Figure 4). It begins just above 60° and proceeds straight northwest through Foxe Channel. Then the bottom of Melville Peninsula is encountered at 66° North. This is just the area that was apparently described in Claudius Clavus's Vienna text. The "swifte ronnyng downe of sea into sea" sounds reminiscent of the *Inventio Fortunatae's* "indrawing channels" and well describes the tidal overfalls of Hudson Strait.

Corte Real was later to become governor of part of the island Terceira in the Portuguese Azores. A local historian would later claim for him credit as the discoverer of lands subsequently visited by explorers from Bristol. Pining later became governor of Iceland.[1]

1474: Paolo Toscanelli. This Florentine doctor of medicine was also interested in astrology, astronomy, and cosmography. He attended the 1439–40 Church Council of Florence, where he met the Portuguese representative, Fernan Martins, of Lisbon. He also met a Greek Orthodox visitor named Plethon, who presumably had been attracted to the event by the recent reconciliation of the Western Church and Eastern Church. This led to thoughts to reattempt establishment of the Far East trade first attempted in the days of Marco Polo. Toscanelli was disturbed by the extreme distance to the Far East, which Polo further emphasized, and by the additional extremities entailed in the Portuguese attempts to find a sea route around Africa. A letter dated 1474 purports to be by Toscanelli. There is some controversy about the authenticity of this letter,[2] but there is no doubt whatsoever of the appropriateness of attribution of these ideas to him at this time.

---

[*]Gemma Frisius's globe of ca. 1537, not reproduced here.

Paul the physician to Fernan Martins, canon of Lisbon, greetings. I was pleased to learn of your good health and of your favor and friendly relations with your very noble and magnificent prince.

On another occasion I spoke with you about a shorter sea route to the lands of spices than that which you take via Guinea [Africa]. And now the Most Serene King [of Portugal] requests of me some statement, or preferably a graphic sketch, whereby that route might become understandable and comprehensive, even to men of slight education. . . . Accordingly I am sending His Majesty a chart done with my own hands in which are designated your shores [the west coast of Europe and Africa] and islands [the recently discovered Azores] from which you should begin to sail ever westward, and the lands you should touch at and how much you should deviate from the pole or from the equator and after what distance, that is, after how many miles, you should reach the most fertile lands of all spices and gems, and you must not be surprised that I call the regions in which spices are found "western" although they are usually called "eastern," for anyone who sails by ship across the under belly of the globe will always find these regions in the West. But if he should go overland by the upper part of the globe, he would come upon these places in the East."

(Then follows a detailed discussion of the wonders of the Orient, as described by Toscanelli's Greek friend Plethon at the Council.) The reference to the "under" and "upper" parts of the globe would be confusing if interpreted in the modern sense of northern and southern hemispheres. Instead, we must face a medieval mappamundi with medieval egocentrism and realize that its lands constituted the upper part of the globe—to us, the Old World hemisphere. The "under belly," to the medieval mind, was the combined Atlantic and Pacific, the Ocean Sea. This attitude was touched upon by the Yale Vinland Map's choice of words to describe the "postreme islands."

Toscanelli was highly respected as a cartographer at the time. The map described here is now lost, but evidently it centered on the Ocean Sea, showing both shores, and was a map of the under part of the globe. Many writers are coming to view the 1457 Genoese/Florentine map of the upper hemisphere as derived from early Toscanelli.[3] Recall that it strongly suggested Norse influence, with the Quebec peninsula making perhaps its appearance as Markland. In fact, Thomas Goldstein endeavored to show a connection between the Yale Vinland Map and Toscanelli. Furthermore, it seems that Toscanelli was familiar with some map by Clavus. During the Council of Florence, Toscanelli showed Plethon a map of the northern part of the Earth made by "a man from Denmark" (as was Clavus). Plethon's later report described it as showing "the great strait opposite the Arctic continent and the land beyond." This mystifying comment will be the subject of further analysis when we come to investigate Clavus's successor Johann Schöner and a possible lost third map by Clavus.[4]

1474–1482: Donnus Nicholaus Germanus and Henricus Martellus. Father Nicholaus was a Germanic cartographer working mostly in Florence who encouraged general usage of Ptolemaic theory by making Ptolemaic atlases. He updated Ptolemy by inventing several new projections. By his editions of about 1467 to 1474 he had begun incorporating modernized geographical data, including a new Clavus-based version of Scandinavia, as appearing in the Zamoiski Codex (Figure 43). The world map from these editions (Figure 44) includes latitudes beyond those on Clavus's chorographic map. It shows that the actual conception of Greenland was as a peninsula reminiscent of the Aleutian peninsula of Alaska.

In a few years, a disciple of Nicholaus, Henricus Martellus, modified the Clavus representation of Scandinavia to account for the knowledge that Scandinavia was not an island. To this purpose he introduced a suggestion of a narrow isthmus, curiously right at the Arctic Circle (Figure 45). Shortly thereafter Martellus modified the isthmus into along narrow neck (Figure 46). Intervening versions by Nicholaus were ambiguous about the isthmus, perhaps intentionally. While the isthmus connection was classic, it had not been used since the development of the portolan more than a century before. Perhaps some of the reluctance to give up the separating strait was based on authoritative information about the reality of Roes Welcome Sound in the hypothesized Southampton Island genesis of this outline of Scandinavia (page 44).

A new edition by Nicholaus himself reflects a complete and fundamental change in the depiction of Greenland. It is embodied in the famous Ulm Ptolemy of 1482 (Figure 47), printed from wood blocks instead of hand reproduced. In this edition the Aleutian peninsula–like prototype for Greenland, with its attachment to the continent at the north of Scandinavia shown in Figures 45 and 46 has been superseded. The wedge shape characteristic of the Quebec peninsula has come back to the north Scandinavian region, under the name Pilapelant. Its relative scale is much smaller than it had been in de Virga and, in fact, smaller than in reality. This reduction in scale brought the western edge of the pictured wedge to he east of Iceland and Scandinavia. This situation has greatly disturbed modern historians, who, with

Nicholaus, erroneously identify the land with Greenland. The Ulm chorographic map, not shown here, labels one corner "Engronelant."

What might have motivated the change to this new depiction? It is interesting to note that both Nicholaus and Toscanelli were working in Florence, and at about the same time. The speculated "third" map of Clavus, which was seen by Toscanelli, could also have been the cause for Nicholaus's change to the new depiction. Scholars have settled on a shorthand way of referring to these two different styles of depicting Greenland. They refer to the earlier style (in Figures 45 and 46) as the A-type and the later style (Figure 47) as the B-type. The speculation is that a lost third map of Clavus motivated the switch to the B-type. (This third map will be further discussed below.)

In association with this Quebec peninsula–like prototype, note an apparent extension of the length of the isthmus to Scandinavia. That is actually brought about by the newly introduced inlet below Pilapelant that is necessary to define its Hudson Bay shoreline. On the other side of the peninsula is an inlet necessitated by the Gulf of St. Lawrence shoreline. This inlet has a *far* shoreline that sweeps northward before turning and arcing southwestward. Such a far shoreline is not necessitated by any graphic requirements, and the map would be artistically better without it. This coastline just might be the vestige of the contorted thinking by Clavus that led to his "Arctic Continent," presumably North America.

The general acceptance of the Ptolemaic scheme of mapmaking, brought about by Nicholaus, was a tremendous step forward for the theory of geography. But it is a decided hindrance in the history we are attempting to trace. While the older mappemonde school had an established tradition of being able through the hiding paradigms to assimilate the contacts of the Norsemen, the Ptolemaic mapmakers did not. Cartographers other than Clavus were initially reluctant to modify Ptolemy's maps anywhere except in regions of which they were certain. The resultant interruption in the Norse cartographic record occurs exactly in the critical decades before the official discovery of America. We shall see that the documentary record continued nevertheless and that the cartographic record later resumed.[5]

1477: Columbus in Iceland. The early adulthood of Christopher Columbus was spent as an ordinary Genoese seaman. That meant bobbing about the coasts of the Mediterranean and participating in the usual adventures of a crew member, while accumulating experience for future leadership. In 1476 one of these adventures took the ship on which he was serving through the Straits of Gibraltar and saw it sunk off Portugal. Being in need of work, he

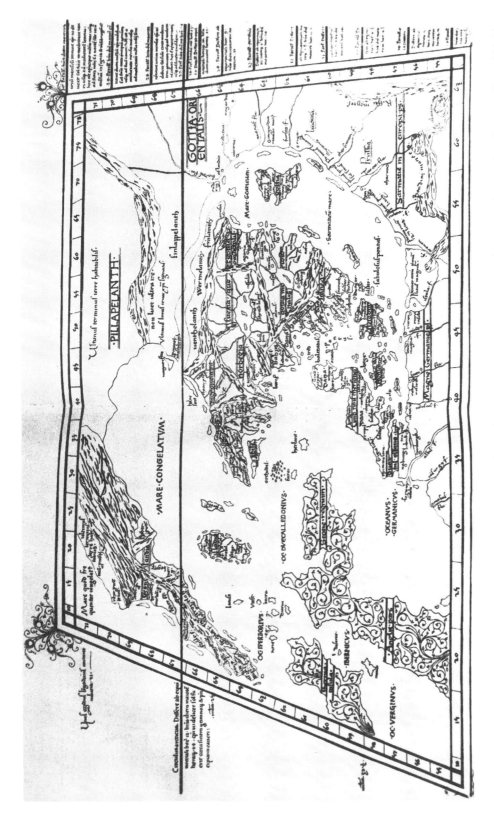

Figure 43. Map of the North by Nicholaus Germanus, in Zamoiski Codex, ca. 1467–74 (representing Clavus, 1431). Location: Warsaw, Zamoiski Library. Source: Nordenskiöld, *Facsimile Atlas*, pl. 30. Courtesy of Map Division, The New York Public Library, Astor, Lenox and Tilden Foundation.

Figure 44. Europe in Nicholaus Germanus's world map of ca. 1467–74. Location: Vatican, urbin. lat. 274, fol. 74. Source: Fite and Freeman, *Old Maps*, p. 4. Courtesy of Map Division, The New York Public Library, Astor, Lenox and Tilden Foundations.

Figure 45. Scandinavia, by Henricus Martellus, ca. 1480. Location: Florence, Biblioteca Mediceo Laurenziana, pl. 29–25. Source: Skelton et al., *Vinland Map*, pl. 16. Courtesy of Biblioteca Medicea Laurenziana and Yale University Press.

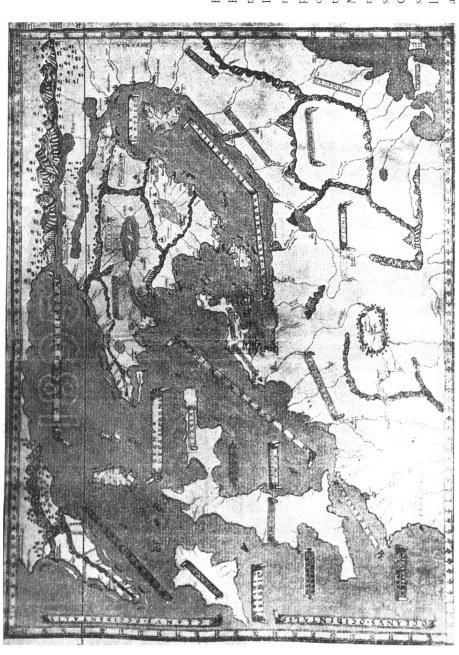

Figure 46. Northwest Europe, by Henricus Martellus, ca. 1481. (Note: Image of "1895" is watermark of paper used for a photoreproduction.) Location: Florence, Biblioteca Nazionale, Cod. Magliabechiano, Cl. 13, no. 16. Source: Nordenskiöld, *Periplus*, p. 87. Courtesy of Map Division, The New York Public Library, Astor, Lenox and Tilden Foundations

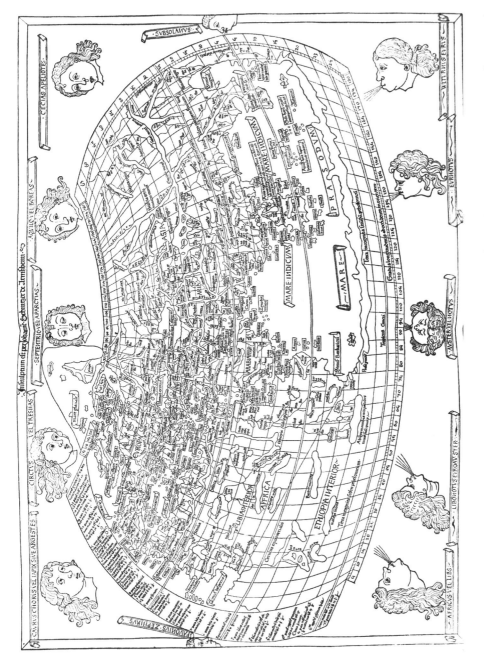

Figure 47. Wood block–printed world map by Nicholaus Germanus (in Ulm Ptolemy), 1482. Location: London, British Library, 1C.9304. Source: Nordenskiöld, *Facsimile Atlas*, pl. 29. Courtesy of Map Division, the New York Public Library, Astor, Lenox and Tilden Foundations.

sought it in the nearest major seaport, Lisbon, where his brother Bartholomew happened to live. The first satisfactory assignment was on the Iceland run. In a note written some years later (according to his son) he said: "I sailed in the year 1477, in the month of February, a hundred leagues beyond Thule, whose northern part is in latitude 73° N. and not 63° as some would have it be; nor does it lie on the meridian where Ptolemy says the West begins, but much further west. And to this island, which is as big as England, come English with their merchandise, especially they of Bristol. And at the season when I was there the sea was not frozen, but the tides were so great that in some places they rose 49 feet and fell as much in depth."

Some writers have suggested that it was during this visit to Iceland that Columbus heard of land in the west. Keeping the source of his information secret, they say, he concocted a plan to sail westward.[6] Certainly the knowledge was generally available without attending any saga-telling parties. That this knowledge reached Columbus seems unlikely, however, for later, when trying to get backing for his project, he went to great lengths to unearth even the slightest scraps of information that would add to the plausibility of his scheme. Knowledge of the Norse explorations could have helped. He evidently did not even firmly conceive of his plans until five or six years later, while on a Portuguese expedition to the Gold Coast of Africa.

Nevertheless, some writers seek to deny the reality of Columbus's Icelandic voyage, not believing the parameters mentioned.[7]* The real significance of both the Icelandic voyage and the African voyage was that now Columbus had seen personally that neither the Frigid Zone nor the Torrid Zone was a death trap. This insight regarding the Frigid Zone would have been heightened by his incorrect belief that in Iceland he had been as far north as 73°. This much error in latitude cannot be blamed on an astrolabe, even given Columbus's infamous error-proneness with the instrument. But he or his son Ferdinand could have computed the latitude from erroneous distance information combined with his famously incorrect calibration of a degree as 57⅔ miles. We have no particular reason to believe he was captain on this trip, but he may have been practicing his navigation. In any case, he learned from personal experience that the greatest danger in sailing off into the unknown was to have insufficient provisions.

Sometime before Toscanelli's death, in 1482, Columbus, his records claim,

---

*Stefansson's reference gives a fairly persuasive argument that Columbus's "*grosse maree,*" traditionally translated as "great tides," should actually be rendered "high waves." Fifty-foot storm waves are not at all unheard of. However, such waves may be encountered anywhere, while Columbus seemed to be reporting something unique to Iceland.

sent Toscanelli a request for detailed information about his ideas. Toscanelli supposedly sent him a copy of his letter and map of 1474. Many scholars (but not all) believe that Columbus's later voyage exactly followed Toscanelli's recipe.[8] Whether Columbus's copy of the letter is authentic or not, there is no doubt whatsoever that Columbus was aware of Toscanelli's ideas through family and business contacts.[9]

1480–1481: John(?) Thloyde and Brasil. This Welsh shipmaster was known to Englishmen as (probably Thomas) Lloyd and operated out of Bristol from 1461. The still-existing customs records show that he sailed to many ports throughout Europe. One curious entry for the year 1466 departs from customs procedure by not naming the destination, simply calling it "exterior parts." It is tempting to correlate this voyage with another of his voyages, recorded imperfectly by a relative of the prominent merchant John Jay of Bristol: "1480, on July 15, the ship [blank] and of John Jay junior, of the burden of 80 tons began a voyage from the port of Kingrode of Bristol toward the Island of Brasil to the west of Ireland ploughing the seas for [blank] and [blank]. Thlyde is the most expert shipmaster of all England; and news came to Bristol on Monday, the 18th of September, that in the said ship they have covered the seas for about nine months [an error for 'weeks'], and did not find the island, but were driven back by storms to the port in Ireland for the restoration of the ship and the men."

It is clear that one objective of the voyage was to reach the historically mysterious Isle of Brasil. That they "did not find" the island is attributed to the storm, not lack of knowledge of the island's whereabouts. The document's original Latin verb *invenio* behind the English translation's "find" has the connotation of discovery, but it also has many other connotations, including that of simply attaining. Even the English concept of "discover" at that time extended beyond original discovery to include exploration and the amplification of original discovery, that is, to dis-cover. After all, the hardnosed businessmen who sponsored these voyages had no use for delusions about an earthly paradise.

Note that the master on these strange voyages had a Welsh name. Many of the pirates operating in northern waters were from Wales. Thloyd was no pirate, but he was in a position to hear reports about their efforts in Greenland and its surrounds. He naturally would then have sought to investigate the economic possibilities of trade. It is worth noting that some of the garments preserved in the graveyards of the Greenland ruins are European styles unique to the latter 1400s. More or less illegal English trade with Iceland and its environs was quite commonplace throughout the fifteenth century. Kirsten Seaver feels it is almost certain that the environs included Greenland, although her evidence is highly speculative.[10]

The subject of trade in a staple product of England, woollens, was naturally sensitive, and even legal voyages were not licensed to trade in such goods. Another such voyage, in 1481, came under suspicion and scrutiny. The only remaining records of the inquiry relate to one Thomas Croft, who was granted this pardon by King Edward IV:

> It is found amongs other thinges that Thomas Croft of Bristoll squier oon of our custumers [customs official] in our said porte of Bristowe aforesaid was possessed of the viij-th [8th] part of a shipp or Balinger called the *Trinite* & of the viij-th part of an other shippe or Balinger callyd the *george* & in to ever of the said shipps or Balingers the said Thomas Croft the same vj day of July the foresaid xxj yere of our reigne [1481] at Bristowe afore said shipped and put xl buschels of salt to the value of xx s [20 shillings] for the reparacion and sustentacion of the said shippes or Balingers and not by cause of marchandise but to thentent to serch & fynde a certain isle called the Isle of Brasile as in the said Inquisicion more playnly yt dothe appere.

Croft had been accused simply of conflict of interest with his customs job, rather than of engaging in unlicensed trade. But there are still some highly irregular aspects of his case. First, the idea that each ship would require forty bushels of salt for the crew's sustenance is questionable no matter long a voyage was anticipated. Second, we must ask whether an inquest would have been called at all unless some destination had been reached to make any allegation plausible. None of the known records says explicitly whether the 1481 voyage reached a destination or not. The king's retrospective reference to the purpose of the voyage does not describe it as simply the search for the island but to "search and find" the island. The Latin phraseology of the inquiry itself to which the king's pardon refers is *scrutando et inveniendo*. This is more suggestive of "exploring and dis-covering" than of blindly groping. About sixteen years later an English informant, John Day, wrote a letter to Columbus about John Cabot's recent activities: "It is considered certain that the cape of the said land [Cabot's landfall] was found and discovered in former times by the men from Bristol who found 'Brasil' as your lordship well knows. It was called the Island of Brasil, and it is assumed and believed to be the mainland that the men from Bristol found." The phrase "in former times" would hardly apply to any discovery in a few years preceding Cabot, but it could apply after a lapse of sixteen years.

The excess salt may help us fill in the blanks in the report of the earlier voyage. Several writers have suggested that the purpose of the salt was to preserve fish, whose locating certainly does require "ploughing the seas." Zeno's fishermen of a century earlier (page 138) may have discovered the Grand Banks fisheries before they were blown to Estotiland. It has been suggested that the efforts of Thloyde and the Bristol merchants were made economically feasible by success in the same area. Perhaps some of the mystery was intentional. Rod and reel fishermen are sometimes willing to share the knowledge of where they are biting, but the Bristol merchants were engaged in a jealous business of making money.[11]

1484: Bergen-Greenland Sailors. In 1647 at Paris, one Peyrere published a book entitled *Relation du Groënland,* which attempted to gather various lesser-known facts about Greenland. One of Peyrere's references was to a Danish manuscript found by the scholar Oluf Worm of Copenhagen. It related an incident that had occurred little more than a century and a half before. The manuscript has not been preserved, but Peyrere's summary of it is that "about 1484, in the reign of King Hans, there were in Bergen, Norway, more than forty experienced men alive, who were in the habit of sailing each year to Greenland and bringing home costly wares. Then in the above mentioned year, some Hanseatic merchants came to Bergen to buy of their wares, but the Norsemen would not do business with them. The Hanse, wishing to avenge themselves, invited the Norsemen to a supper, and after mealtime killed them all that same night"[12] Peyrere used this story in a forced argument to show why people no longer knew the sailing directions to Greenland. The story has the fictitious ring of being a too-perfect example of the legendary Hanseatic harshness. Other than this exaggeration, however, there is no reason to doubt that the story grew out of some actual incident in the year specified. The information that there were annual voyages to Greenland before 1484 was not crucial to Peyrere's argument. Only the suggestion that after 1484 there was nobody left who knew the route mattered to Peyrere. Such incidental information is frequently more to be trusted than the explicit message of an argument. Whoever wrote the original manuscript obviously felt no surprise that annual voyages were being made for some time up to 1484. If so, there must still have been somebody there in Greenland.

1486–1487: Ferdinand van Olmen. This Flemish settler in the Azores was governor of the northern half of Terceira simultaneously with João Vaz Corte Real's governorship of the southern half. In March of 1486 the Portuguese king, João II, issued the first of several exploration patents to van Olmen. All of them related to an Atlantic voyage projected for March of 1487. The plans for this voyage were similar to many others that had tried to identify the unknown islands on the nautical chart of 1424 with the Isle of Seven Cities in a Portuguese legend from the Dark Ages. However, van Olmen's plans differed from those of the other island hunters in that he proposed "to seek and find a great island or islands *or the coast of the mainland*" (emphasis added). In this context *mainland* would have meant Orbis Terrarum. That van Olmen's interest lay towards the northwest is confirmed by the nearly contemporary historian Las Casas, who wrote of a contemporary Portuguese voyage, "During the Ireland run they were heading so far to the northwest that they saw land to the west of Ireland, which they believed must be that which Ferdinand van Olmen sought to explore."[13] There remains no proof of whether or not van Olmen's expedition ever actually sailed, but he did have strong and level-headed backing. If João Vaz Corte Real actually did participate in the voyage of over a decade before (page 195), he would have encouraged this one to the same vicinity. Not just deeds, but ideas, are important to this history.

1491–1498: Bristol Voyages Renewed. The historical record is so far still mute on the "exploratory" activities of the Bristol merchants from 1482 to 1490, but in 1498 the Spanish ambassador Pedro de Ayala wrote from England to his sovereigns, "For the last seven years the people of Bristol have equipped two, three and four caravels to go in search of the Island of Brasil." This was written before the known departure of a 1498 voyage, thus placing the first of the seven annual voyages in 1491. The risking of two, three, and four caravels each year could not have been justified without some economic return. That the voyages were still considered exploratory rather than established may mean that the merchants were concealing their activities from someone. It seems likely that the names we have seen in the forefront—Thloyde, Croft, Jay, and later, Cabot—formed an inner circle in Bristol who shared information among themselves.

Another mystery is the silence in the record from 1482 to 1490. It seems evident that the new voyages were an extension of the activities from the time when Thloyde was England's master mariner. The silence may have merely been a discreet recognition of political realities. From 1484 to 1490 England was at war with Denmark (i.e., Scandinavia) and the Hanseatics. However, the peace treaty of 1490 gave England a new, previously unheard of legal privilege—the right to trade directly with Iceland, and presumably with related islands.[14]

1492: Bishop Mathias Knutsson. In 1492 a new pope, Alexander VI, was elected. Sometime during the first year of his papacy he issued the following document:

> As we are informed, the church at Gardar lies at the world's end in the land of Greenland, where the people for want of bread, wine and oil, live on dried fish and milk; and therefore, as well as by reason of the extreme rarity of the voyages that have taken place to the said land, for which the severe freezing of the waters is alleged as the cause, it is believed that for eighty years no ship has landed there; and if such voyages should take place, it is thought that in any case it could only be in the month of August, when the same ice is dissolved; and for this reason it is said that for eighty years or thereabouts no bishop or priest has resided at the church.* Therefore, and because there are no Catholic priests, it has befallen that most of the parishioners, who formerly were Catholics, have (oh how sorrowful) renounced the holy sacrament of baptism received from them; and that the inhabitants of that land have nothing else to remind them of the Christian religion than a corporal [eucharistic altar cloth] which is exhibited once a year, and whereon the body of Christ was consecrated a hundred years ago by the last priest who was there.
>
> For this reason as well as other considerations, our well remembered predecessor (among whose descendants we have now been appointed), Pope Innocent VIII, wishing to provide them first with the word of the Church and next with the lost pastoral comfort of a fitting and proper shepherd, has out of the brothers' own council [the Curia or College of Cardinals?] (one of whose number we then were), put our honored brother Mathias of the order of St. Benedict chosen Gardar [i.e., bishop of Gardar], the same having volunteered out of respect for our presence to take up, according to announced ecclesiastical policy and great devotional ardor, the office of *accensus* in the face of rumors of deviators and apostates—has put the same bishop and shepherd in charge of such restorings to the way of eternal life and rooting out of errors, he freely and of his own accord offering up his life to the very largest danger to set out making hence personally by

---

*Several had been appointed but had remained in Europe.

ship. We therefore of the same choice have commissioned and do mandate the conscientious and praiseworthy plan both in God and in having sent in advance to other places the strongest benefices for assistance of rescue from his poverty of means, the existence of which we have been told more reluctantly, to come to the help of needs, in personal feeling and also in accordance with our definite knowledge from the council and approval of the brothers of ours, the place for replying to the beloved sons [the Curia], that whoever of our apostolic officials and those keepers of registry lists and others as much the Chancellery as the Treasury who would erect obstacles or trouble or delay, by the very nature of the deed immediately running into provisions for the blanket sentence of excommunication, considering how each and every apostolic document about and concerning the commanded travel to the church of Gardar, according to the dictum "Select" arranging for the services of each and every one of these for nothing, should according to edict be settled and made free without tax, payment or assessment of any kind, anything to the contrary notwithstanding. Also the apostolic Treasury, clerks and secretaries, considering how they should deliver and register such dispatches or bulls in accordance with the dictum "Select" without payment or assessment of anybody's annates or similarly of petty servitude and of any other rights whatever, are to be released freely of customs; with similar feeling and knowledge and under the provisions beforesaid we have commissioned and do mandate that, making no obstructions whatever, one should on the contrary function for nothing everywhere because of the poor, etc.

Dated: Anno Primo[15]

Written in florid Latin, this document gives a glimpse into the decline of Greenland. However, it should not be taken at face value. The statement that the rarity of the voyages was caused by the "severe freezing of the waters" does not sound like a first-hand report but like a reasoned deduction of robed scholars. The idea that one could only sail in August is completely out of touch with reality.

The statement, "It is believed that for eighty years no ship has landed there," conforms roughly with the sack of the Eastern Settlement by pirates in 1418, recorded in Pope Nicholas V's letter of 1448. The present letter and proposed expedition could have been motivated as a follow-up. Over the previous century many bishops had been appointed to Greenland but never went there. They were not really expected to go. Seaver thinks this was another such situation, and that the new bishop was just a "confidence man."[16]

The document does give Mathias great power, which he could abuse, if he intended to. However, why would such a person—not just another priest

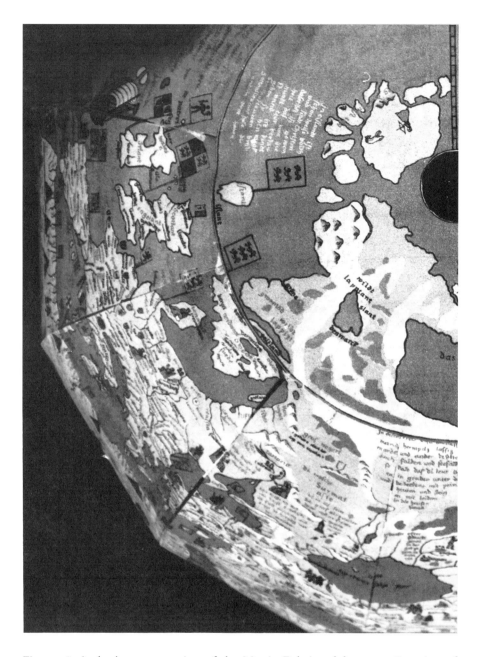

Figure 48. Author's reconstruction of the Martin Behaim globe, 1492. Location of original: German National Museum, Nuremberg. Source: E. G. Ravenstein, *Martin Behaim: His Life and His Globe* (London: 1908), rear pocket gores.

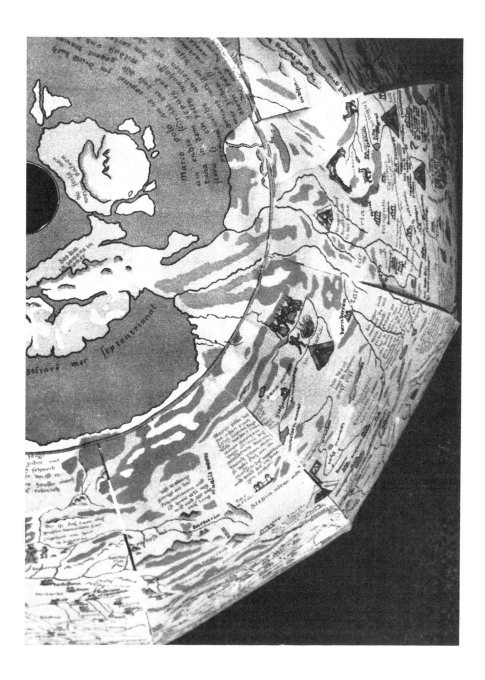

but a member of the Curia—want essentially to become a vagrant? Perhaps he had information that gave him hope that he really could still find passage to Greenland.

1492: Martin Behaim's Globe. The Behaim globe (see Figure 48) has been discussed extensively earlier in this book. It probably contains the Arctic Archipelago, the neighboring Canadian Arctic coast, and Hudson Bay. Controversy about whether Behaim himself actually drew the globe is irrelevant here. He certainly was aware of and responsible for whatever went into it. Martin Behaim was a Nuremberg mathematician who moved to Lisbon in 1484. There he is thought to have been appointed to a *Junta dos Mathematicos* of Portugal. This was an ad hoc royal maritime advisory committee, organized at intervals to assist Portugal's explorations in Africa. Its activities were designing new navigational techniques and instruments as well as collecting general geographical knowledge.

When Columbus first submitted his "Enterprise of the Indies" to the Portuguese king in 1484, the members of such a committee rejected the plan as unfeasible. They held that it violated known theory on the size of the globe and land. The exact date of Behaim's membership in the committee and whether he was among those who considered Columbus's proposal are unknown. Behaim presumably would have been voting on the other side, in favor of the proposal. It is not certain if Columbus and Behaim ever met one another. Nevertheless, a body of indirect evidence suggests that they may have at least been aware of each other's efforts. They had identical small estimates of the size of a degree, and Behaim later desired to undertake the westward voyage after Portugal had rejected Columbus's scheme a second time. Behaim's overall coastlines and land-to-sea ratio are similar to those of a Henricus Martellus map of ca. 1489 (not shown here), the parameters of which are believed by some to have inspired Columbus.[17] We can see rather convincingly from Behaim's globe that his belief in a narrow ocean had strong corroboration from information based on human experience in North American lands.

SUMMARY

Even though European contact with Greenland was supposed by previous scholars to have ceased by 1420, during the 1480s and 1490s at least six relatively independent movements for exploring north and west appeared. Something seems to have been "in the air." The flurry of planned as well as executed voyages westward includes several that seem clearly motivated by information originating with or transmitted by the Norse. It would be unrealistic not to consider seriously the possibility that in fact all such westward attention was so motivated, directly or indirectly.

CHAPTER 13

MASTERY OF THE ATLANTIC

THE NORSEMEN REACHED AMERICA BY A SERIES OF FORESHORTENED STEPS between islands in the high latitudes. Even so, the duration of many of those steps was measured in generations. For southern Europeans to become familiar with America required developing the ability to sail straight across the Atlantic Ocean and back again with reasonable assurance of success.

1492–1498: Columbus in the Caribbean. If Leif Erikson was lucky, Columbus was blessed by Fortune herself, for if the islands of the West Indies had not obstructed his way "to Asia," we never would have heard from him again. His provisions and crew would probably not have lasted even to the Central American mainland.

No one has ever doubted that Columbus attained South America (although not until 1498), and he did trace along Central America in 1502. But no scholar of history has ever claimed that he did discover North America. His real contribution was to prove the reliability of the Atlantic trade winds, which had been discovered in previous decades by the Portuguese and others exploring for islands.[1] The question "Who discovered the New World, Columbus or the Norsemen?" is specious, posed only by some journalists looking to create news. The "New World" was actually "discovered" by theoreticians in Europe. They never even left their armchairs while they put together the two's and two's of many explorers tales. Even then, it was not the New World as we see it. The first six years after Columbus's long voyage were spent exploiting the islands of the Caribbean, and during that time he came nowhere near mainland. His own faith and his interpretation of native accounts led him to believe that the mainland of Asia was nearby. He convinced the general Spanish public that he had accomplished all he had set out to do.

But the geographers knew the impossibility of this. They placed the discoveries as simply more islands out in the ocean. The royal Spanish backers of Columbus encouraged him simply to develop the riches of these islands. Outside of Spain, Portugal, and Italy, little significance was attached to the discoveries initially. Everybody, however, marveled at the bravery and skill of the long voyages.[2]

1493: Hieronymus Müntzer Letter. Hieronymus Müntzer can be thought of as a German counterpart of the Florentine Paolo Toscanelli. Where the Genoese Columbus supposedly sought Toscanelli's reassurance in presenting his plan to King João II of Portugal, Müntzer proposed to the king a similar voyage by his fellow Nuremberger, Martin Behaim. When Müntzer wrote his endorsement in July of 1493, he either had not yet heard the news of Columbus's return in March or, perhaps more likely, discounted its significance.

To the Most Serene and Invincible João, King of Portugal, the Algarves and Maritime Mauritania, first Discoverer of the Islands of the Canaries, Madeira and the Azores, Hieronymus Müntzer, a German doctor of medicine most humbly commends himself. Because heretofore you have inherited from the most serene prince Dom Henrique your uncle the glory of sparing neither effort nor expense in extending the bounds of the world and by your diligence you have subjected to your rule the Seas of Ethiopia and Guinea and the maritime nations as far as the Tropic of Capricorn together with their products such as gold, grains of Paradise, pepper, slaves, and other things, you have by this display of talents won renown and immortal fame and great profit for yourself besides. . . . Wherefore, Maximillian, the most invincible king of the Romans [the Holy Roman Empire, essentially Germany], who is Portuguese on his mother's side, desired to have Your Highness invited to seek the very rich shore of Catay in the Orient by way of a letter, however crude, written by me. . . . [It is my opinion] that the beginning of the habitable East is quite close to the boundary of the habitable West. . . . Moreover, there are arguments without number, so to say, and of great persuasiveness, for concluding almost without question that this sea can be crossed to the Catay of the Orient in a few days. . . . For in matters pertaining to the habitableness of the earth more confidence is to be placed in experience and trustworthy accounts than in fantastic imaginations. . . . Already you are praised as a splendid Prince by the Germans, Italians, Russians, Polish, Scythians, and those who dwell beneath the arid star of the arctic pole, as well as by the Grand Duke of Muscovy—and not many years ago under the aridness of the said star the great island of Greenland was discovered,* which has a circumference of 300 leagues and in which there is a large settlement under the said sovereignty of the said Grand Duke. But, if you carry out this mission you will be exalted with praise like a god or a second Hercules; and, if you wish, you will take along on this journey as a companion Martin Behaim, deputed by our king Maximillian especially for this purpose, and many other learned mariners who will set out from the Azores and sail the breadth of the sea, by their skill, with the aid of the quadrant, the chilindre, the astrolabe, and other instruments.[3]

While Müntzer signed the letter, one might expect that Behaim was actually the instigator. Behaim had just recently finished his globe, which, as we saw, contained information from the *Inventio Fortunatae*. Through the Arctic references, one can almost hear Behaim's sources crying for recognition in this letter. The urging by Maximillian makes one wonder if these sources may not have also been tied up with the Hanseatic League. The proposed structure of the expedition is closely modeled after the Scolvus voyage, which the Germans presumably learned about when the Hanseatic League took control of the primary Scandinavian ports.

1497 and 1498: Encounters with America. During these years there were several independent contacts with mainland in both North and South America.† These contacts included those by John Cabot, Americus Vespuccius, and Christopher Columbus. Vespuccius's and Columbus's accidental contacts in South America were clearly extensions of Columbus's earlier explorations in the Caribbean, but the activities of Cabot in North America are not at all clear. The most likely interpretation of the evidence at present is that Cabot (Giovanni Caboto) a land holding and educated Venetian, was in Spain in 1493 when Columbus returned from his first voyage amid great publicity. Examining the evidence first-hand Cabot decided that Columbus had not actually reached Asia. He was nevertheless fired by the realization that such long voyages out of sight of land were humanly possible. He developed his own desires to accomplish what he felt Columbus had failed in. Somewhere

---

*The name Greenland has been applied to many a discovery throughout history. In this case it does not apply to Erik the Red's island, but to Spitsbergen, which was discovered by Russians in 1435 and briefly settled under the name Greenland.

†Explorers determine mainland not by its continuity with other land but by the characteristics of its rivers, flora, and fauna. Thus, a relatively short exploration can clearly identify mainland. Of course, large islands can offer misleading evidence, as Cuba did to Columbus.

along the way he sought and was refused assistance in his own project at Seville and Lisbon. Eventually he found himself in Bristol. There he must have heard about the voyages to the Isle of Brasil and attached to them a significance that had escaped the less theoretically inclined merchants. He may have reasoned that the distance from Europe to Asia along the latitude of Columbus's route must have been far greater than Columbus's actual sailing distance but that the distance in latitudes nearer the top of the globe would be less. Furthermore, the unknown northern part of Asia beyond Marco Polo's Orient may well project still farther eastward and provide a coastline to follow southwest to the riches of Cathay. Perhaps the Bristol merchants had already discovered the easy passage to Asia, without knowing it, he may have thought. Cabot meant to find out, and he evidently persuaded the merchants to allow him to take charge of the voyages from 1496 onward. While the 1496 voyage was unsuccessful, the 1497 voyage reached mainland, and shortly after its return, Columbus's informant John Day wrote to his lordship about Cabot's voyage.

> They spent about one month discovering [exploring] the coast and from the above mentioned cape of the mainland which is nearest to Ireland, they returned to the coast of Europe in fifteen days. They had the wind behind them, and he reached Brittany because the sailors confused him, saying that he was heading too far north. From there he came to Bristol, and he went to see the King to report to him all the above mentioned; and the King granted him an annual pension of twenty pounds sterling to sustain himself until the time comes when more will be known of this business, since with God's help it is hoped to push through plans for exploring the said land more thoroughly next year with ten or twelve vessels—because in his voyage he had only one ship of fifty toneles and twenty men and food for seven or eight months—and they want to carry out this new project. It is considered certain that the cape of the said land was found and discovered in former times by the men from Bristol who found Brasil as your Lordship well knows. It was called the Island of Brasil, and it is assumed and believed to be the mainland that the men from Bristol found.
>
> Since your Lordship wants information relating to the first voyage [1496], here is what happened: he went with one ship, his crew confused him, he was short of supplies and ran into bad weather, and he decided to turn back.

These passages leave little doubt that Cabot was following rather than leading his crew, who evidently were selected because of their prior experience sailing to those regions. The phrase "as your Lordship well knows" also leaves little doubt that Columbus and the Spanish were for some time keenly aware of the Bristol activities. Since he was responsible for rescuing the significance of the discovery, Cabot accepted credit for the discovery itself. A contemporary report says, "He is called the Great Admiral and vast honor is paid to him and he goes dressed in silk, and these English run after him like mad."

Funding was acquired for a further grand expedition which departed in 1498. Cabot had taken over the glory that Columbus had experienced upon his return from his first voyage. Furthermore, in Cabot's case there was no doubt that mainland had been reached. However, a very important semantic distinction must be made here: this was *not* a discovery of *America* in a strict sense; the continent reached was still thought to be Asia (Orbis Terrarum), and the realization of a separate America would take at least another decade.

Meanwhile Columbus was having more and more difficulty convincing the world that his own discoveries were not just some islands out in the ocean which he had trouble administrating. Columbus was later destined to suffer the indignity of being taken back to Spain in chains, under the arrest of a new governor of the Indies who had been appointed by the sovereigns to supersede him. Before this embitterment, while exploring the island of Trinidad in 1498, he did encounter mainland at the Paria Peninsula of South America (shown in Figure 49). Even this discovery, however, was not to reinstate his lost glory, for he was torn between identifying it as mainland or identifying it as an island he called Gracia. In summarizing Columbus's now-lost log book, a contemporary historian, Las Casas, wrote from Hispaniola (Haiti), "He says that the whole sea is sweet [fresh water] and that he does not know whence it springs, for it did not seem to be fed by large rivers, and he says that even if it had it would not cease to be miraculous." What they had come close to discovering was one of the mightiest rivers in the world, the Orinoco; but even after sending a caravel to investigate the interior of the Gulf of Paria, Columbus did not accept the sailors' report: "The Admiral thought that these four outlets or openings were four islands and that there did not appear to be any sign of the river which made that whole gulf of 40 leagues of sea, all sweet water; but the sailors asserted that the openings were mouths of rivers." In the end, not having gone ashore himself and having no time to explore farther before he was due back in Hispaniola with supplies, Columbus stuck to his island theory, for "he knew that neither the Ganges, nor the Euphrates, nor the Nile carried so much fresh water. The consideration which moved him was that he did not see lands large enough to provide a source for such large rivers, unless, he says, this land is a continent."

Even though all of his hopes had been tied up in reaching Asia and he was certain that Cabot had already done so, Columbus could not believe that this land was part of Orbis Terrarum. He knew that the coast of Asia must lie west of his islands, and this land was south and east of them. However, on the trip back to Hispaniola he had time to do some deep thinking about the matter, and eventually made the following entry in his log: "I have come to believe that this is a mighty continent which was hitherto unknown." While it was hitherto unencountered, Columbus nevertheless believed it to be a continent that had been predicted by geographical theoreticians. This was a type of antipodal continent, as exemplified in the so-called Zorzi sketch map (Figure 49) originated by his brother Bartholomew in 1503 after part of the coastline had been surveyed.* On a companion map, Bartholomew actually labeled this continent "Antipodi."[4] Even those who did not share Columbus's idea of the distance to Asia nevertheless believed in the theoretical existence of a western passage to Asia north of the Antipodes. It was partly the search for this passage that led to a complete discovery of the eastern seaboard of the Americas, but this operation required many decades. Recall Albert von Szent-Gyorgy's definition: "Discovery consists of seeing what everybody has seen and thinking what nobody has thought." In this sense it is no more correct to say that Columbus discovered America than to say that the Norsemen, whose North American encounters appeared in Scandinavian and northeast Asian maps, discovered America. Columbus considered the new continent merely a diversion from his goal of Asia. The Florentine Americus Vespuccius paid timely attention to the continent, however, and popularized the term *New World,* but not in the sense that we assign to those words today. He was to be rewarded (although amid controversy) by having the continents of the New World named after him.[5]

1500: Juan de la Cosa. There may have been two people by this name associated with the same part of history. One was the owner of the *Santa Maria* and the other an able seaman on the later voyages of the *Nina* who had a reputation as a chartmaker. Many scholars think they were the same man. In 1500 the latter La Cosa drew a world map (Figure 50) which, since its rediscovery in a Paris antique shop in 1832, has been the subject of more speculative comment than any other map, including the Yale Vinland Map. The

---

*This, as well as other sketch maps by Bartholomew, appeared in the margins of a copy of Christopher's famous "Letter," published by Alessandro Zorzi after the first voyage.

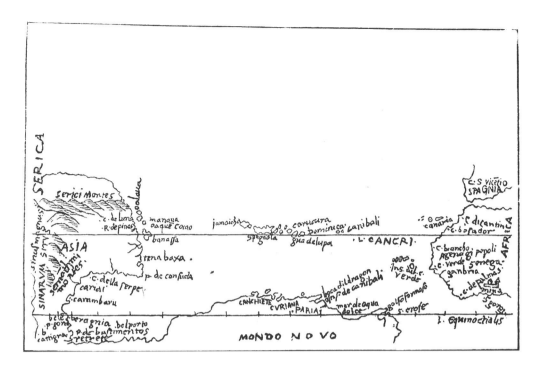

Figure 49. Sketch map by Bartholomew Columbus, 1503. Location: Florence, Biblioteca Nazionale, Classe 13, Cod. 81, fol. 31–34. Source: Nordenskiöld, *Periplus,* p. 163, fig. 79. Courtesy of Map Division, The New York Public Library, Astor, Lenox and Tilden Foundations.

southern part of La Cosa's chart of the new hemisphere prompts little controversy; it is based clearly on known explorations of the antipodal coastline. Just north of it is an allegorical representation of Columbus plumbing the waters for the Western Passage to Asia, carrying Christianity to the new lands.

The northern part of the map has caused controversy. It seems on first glance to represent the entire seaboard of North America. Legends and English flags along part of this coastline (the "Named Coast"—of controversial legibility) have been presumed to refer to the 1497 explorations of John Cabot. The rest of the coastline is presumed to be from Cabot's 1498 voyage. However, there is no evidence whatsoever to corroborate this assumption, and we shall see the existing evidence point in an opposite direction. First, aside from this map there is no evidence that John Cabot ever returned from his exploratory expedition of 1498, the only one on which such a map could have been surveyed. The ships were subjected to a damaging storm on the outward passage, from which one limped back with the news that Cabot was persisting anyway. This follow-up expedition had departed amid great fanfare, but even contemporary historians noted the lack of evidence of his

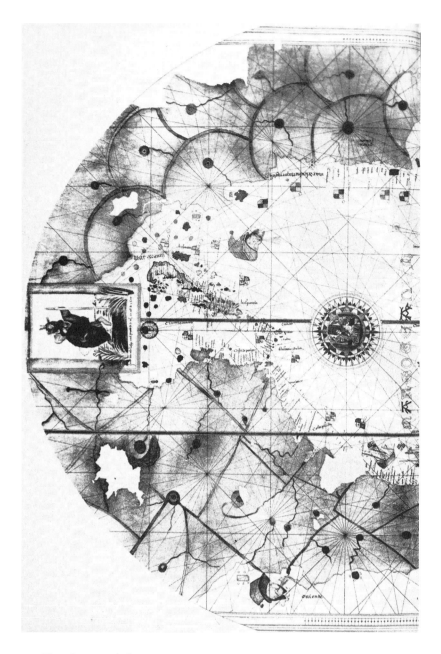

Figure 50. Chart by Juan de la Cosa, 1500. Location: Madrid, Naval Museum. Source: Nordenskiöld, *Periplus*, pl. 43. Courtesy of Map Division, The New York Public Library, Astor, Lenox and Tilden Foundations.

having returned alive, suggesting strongly that the expedition came to grief. The claim of La Cosa's map as being evidence for Cabot's return is circular reasoning, since the return of Cabot was invoked in the first place to explain the La Cosa map. A discourse by a Mantuan gentleman in 1550 states that

Cabot's death coincided with Columbus's encounter with the "coast of the Indies" (the mainland), which would have been in 1498.[*]

Ian Wilson has invented a scenario taking the remaining 1498 Cabot ships

---

[*] The Mantuan gentleman's statement was supposed to be a quotation of Cabot's son Sebastian, who apparently allowed the statement about the timing of events to be misunderstood as referring to Columbus's 1492 discovery. This had the effect of making some historians give Sebastian credit for his father's activities, an effect evidently pleasing to Sebastian.

through the storm and down the American eastern seaboard all the way around the Gulf of Mexico and Caribbean Sea to Venezuela. There they hypothetically encountered a Spanish voyage under Alonso de Hojeda which included La Cosa, and Cabot's crews imparted the new knowledge to La Cosa before they were executed.[6] This conspiracy-laden fantasy is easily discredited. Wilson cites the modern Spanish historian Martin Fernandéz de Navarrete's unsubstantiated surmise, "It is certain that Hojeda in his first voyage [that of 1499] encountered certain Englishmen in the vicinity of Coquibaçoa." Wilson claims that this is supported by words in Hojeda's patent for a 1501 voyage: "that you go and follow the coast which you have discovered, which runs east and west, as it appears, because it goes toward the region where it has been learned that the English were making discoveries; . . . so that you may stop the exploration of the English in that direction." However, Navarrete's statement is not "supported" by this, but probably *based* upon it. The concern about English explorers in the Spanish sphere is sufficiently explained by Ayala's letter (page 209), and the wording of the patent makes it clear that the concern was preventive rather than reactive. Whatever was the prototype for La Cosa's map, he naturally associated it with the English expeditions he had heard about. But that prototype is not likely to have resulted from a survey of North America as we know it, from Labrador to Florida. This brings us to the second problem in the interpretation of the map.

La Cosa's map is, no matter what kind of allowances are made, a rather poor representation of any section of the eastern seaboard of North America. Its primary likeness to North America is that it has been placed by La Cosa in the northern part of the Western Hemisphere. A host of projections and distortions have been invoked to explain why the coastline tends eastward instead of northward. None of them is quite successful, and the peninsulas, islands, and inlets all occur at the wrong relative places and in the wrong proportions. Furthermore, La Cosa himself has bathed the entire land mass in a green color that he uses elsewhere on the map to indicate uncertainty about his source.[7]

There is, now, a third point to be considered. Yes, it is doubtful that Cabot ever returned with a survey of North America, even from his 1497 voyage. It is nevertheless clear that by some unknown provenance he had some map of a western region before he departed in 1498, and conceivably even before 1497, a map which was to guide his exploration. A London chronicle has in part the following entry for the year 1498: "This yere the kyng at the besy request and supplicacian of a Straunger venisian [Cabot], which by a Caart [chart] made hym self expert in knowyng of the world, caused the kyng to manne a ship with vytaill & other necessaries for to seche an Iland wheryn the said straunger syrmysed to be grete comodities."

The island being referred to was, of course, the Isle of Brasil. The chart referred to is not likely to have been any of the standard maps since 1325 which showed this island. For Cabot to receive the audience to make himself expert before the king he needed a unique map and one with convincing credentials. It is evident that copies of this now unknown map were made and distributed. Shortly after the 1498 expedition had departed, a Spanish ambassador in London wrote home to his sovereigns:

> I think Your Highnesses have already heard how the king of England has equipped a fleet to explore certain islands or mainland which he has been assured certain persons who set out last year from Bristol in search of the same have discovered. I have seen the map made by the discoverer, who is another Genoese like Columbus,* who has been in Seville and at Lisbon seeking to obtain persons to aid him in this discovery. . . . I told [the English king] that I believed the islands were those found by Your Highnesses, and although I gave him the main reason, he will have none of it. Since I believe Your Highnesses will already have notice of all this and also of the chart or mappemonde which this man has made, I do not send it now, although it is here, and so far as I can see exceedingly false, in order to make believe that these are not part of the said islands.[8]

These passages clearly must refer to the map made *before* the 1498 departure, in order for him to assume that knowledge of it had already reached Spain. It is entirely possible that it was one of the copies of that widely circulated map that found its way into La Cosa's map, perhaps in a distorted form. It would then be possible that the coastline in the original Cabot prototype came nowhere near the West Indies. To justify their claims to the land, the Spanish may have "corrected" the "exceedingly false" prototype to the form in La Cosa's map. Yes, these are speculations, but the map's attribution to Cabot is also pure speculation. We will cast our speculation as an auxilliary hypothesis, later to be supported by evidence more convincing than any the Cabot theory has marshaled. There will be evidence from 1506, a map drawn independently of Spanish influence but from the same or a related prototype.

1501: Piri Reis's Informer. In 1501 several warships from the Ottoman Empire engaged several Spanish vessels near the shores of Valencia and overcame

---

*Cabot was a naturalized Venetian, but his place of birth is thought to have been Genoa.

seven of them. There were many Spanish captives. It was probably one of these, who had by coincidence served as a crewman on some of the first three voyages of Columbus, who was made a slave to the Turkish admiral. The Turkish admiral's nephew Piri, himself later an admiral, incorporated the captive's story about Columbus (and a captured map) into his own later maps and navigation books. The story thus preserved, accurate in many verifiable respects, includes this account: "a Genoese infidel, his name was Colombo, he it was who discovered these places. For instance, a book fell into the hands of said Colombo, and he found it said in this book that at the end of the Western Sea, that is, on its western side, there were coasts and islands and all kinds of metals and also precious stones." The Spanish informer was captured *before* there was any controversy about Columbus's inspiration, and this story can be taken at face value.

One might speculate that the unnamed book describing the riches of that coast was Marco Polo's *Travels.* It is known that that book was important to Columbus. However, that speculation might be jumping to a conclusion, because Marco Polo never referred to the Western Sea and had no western orientation whatsoever. Pierre d'Ailly's *Imago Mundi,* sometimes suggested as this book, did have a westward orientation. Another candidate has been the *Historia Rerum Ubique Gestarum* by Aeneas Sylvius Piccolomini (Pope Pius II). Yet another possibility arises from the fact that in his *Bahriye,* Piri Reis refers again to the book, in terms that might indicate Ptolemy's *Geographia.*[9]

A further book that Columbus is known to have read also had a westward orientation: the *Inventio Fortunatae.* If the book being referred to was indeed the lost *Inventio,* then we have additional information about its contents, namely, that it referred to the mainland coast of America as well as the Arctic islands that surviving traces appear to have described. Indeed, it might have described the coast as far south as the Mexican cultures in order to refer to precious metals and stones. We know from statements by Ferdinand Columbus and Bartolomeo Las Casas that the *Inventio* did in fact cover as far south as a latitude opposite the Cape Verde Islands.[10] That would be 18° N, the exact latitude of the Caribbean islands to which Columbus first sailed.

1500–1502: Azorean Explorations. The Portuguese nobleman João Vaz Corte Real, who supposedly accompanied the poorly understood expedition of Pining, Pothurst, and Scolvus to the Greenland area in 1472, subsequently settled down as Portuguese governor of the southern half of Terciera in the Azores. During the same period Ferdinand van Olmen, who had sought the mainland in 1486–87, was governing the northern half. In 1500 Gaspar

Corte Real, the son of João Vaz, apparently discovered Newfoundland after several poorly documented previous expeditions to the area. It seems reasonable to suspect that these expeditions were sparked by information from João Vaz's and van Olmen's memories. Gaspar's flagship never returned from a two-ship 1501 repeat voyage, and his brother Miguel took over for a 1502 voyage. Miguel's flagship also never returned, but two other ships did. A third brother, Vasco Annes, was royally prohibited from further pursuit, after a royal search expedition for the missing two was unsuccessful. Nevertheless, the Portuguese road to Newfoundland had been opened officially, without any obfuscation.[11]

SUMMARY

The discovery of America proceeded rather quickly, even though with many difficulties, once European ships had the confidence to cross the ocean directly. But this could have been merely the last chapter of a process that was spread over more than two centuries. Numerous mysterious threads pervade the history of the Bristol activities, suggesting a tie to the Norse record.

OLD IMAGES IN NEW MAPS

CHAPTER 14

A NEW CONTINENT EMERGES

WE HAVE SEEN THAT EUROPEAN CARTOGRAPHERS STRUGGLED for centuries with a wide variety of anomalous data from Norse sources, using whatever paradigms they had to incorporate it into their maps. This source data was often oral, but they may also have made hard copies of their source data or some of it may have come as hard copy. None of their source data has survived into the twenty-first century, perhaps understandably. But some of it quite possibly could have survived into the sixteenth century. An unidentifiable map from some Scandinavian source could have been put on a back shelf instead of thrown away.

The European exploration of the Western Hemisphere naturally started a completely fresh era in the development of the world map. European cartographers had to reject completely the restrictive traditions of medieval mapmaking, and they had to become eager searchers after every scrap of geographical hearsay. In the process, some of them might have remembered that strange map on the back shelf and found a place on their new map where it seemed to fit. This would have been latent foreknowledge showing up after the discovery of America. We will see numerous examples of this below.

As a whole new generation of geographical theoreticians turned to the problem of making sense out of the reports produced by dozens of new explorations, a strange inconsistency arose in their work. The cartography of North America required over half a century to throw off these confusions, while the cartographical picture of South America developed quite rapidly into its correct form. Some of this difference is attributable to the widespread theoretical conclusion that South America was a new world, the Antipodes, being newly mapped, while North American land was still considered Asian. Part of the inconsistency is also doubtlessly explained by the national diversity among the explorers of North America. Another reason, we will see, is that Norse-derived maps continued to interfere in the geographical conceptions of the North without affecting the appearance of South America on the world map.

1502: Cantino Map. The so-called Cantino map of 1502 (western portion shown in Figure 51) was drawn by an unknown Portuguese cartographer (and/or a Flemish cartographer in the employ of the Portuguese) and delivered to its Italian purchaser via one Alberto Cantino. It shows in the New World a tree-covered "land of the King of Portugal." This represents the southeastern part of Newfoundland as surveyed by the vessels which returned from the Corte Real expeditions. Later maps showing the same land call it "Terra Corterealis." However, the map also contains an amazingly accurate survey of the southern part of Greenland. A scroll attached to Greenland says, "This land was discovered by license of the most excellent Prince D. Manuel King of Portugal, and they who discovered it went not ashore, but viewed it and saw nothing but very thick mountains, whence according to the opinion of cosmographers it is believed to be the peninsula of Asia." The "peninsula of Asia" would refer to Pliny's Tabin Peninsula, Asia's supposedly ultimate northeast peninsula. "Went not ashore," indeed! To survey this vast coastline with such accuracy, even including the myriad islands along the west coast, would have been so formidable a mission that they would have had no time left to survey Newfoundland. They would have had to spend every second available at sea under full sail without making a single navigational error while negotiating the unknown coastlines. Even then the fall drift ice would have caught them before they got very far. It is highly improbable that anybody could have sighted the ice cap of Greenland and seen nothing to report but "very thick mountains." Perhaps more plausibly, this map of Greenland came from another unknown Norse or Eskimo prototype instead of from contemporary Portuguese explorers. Modern expectation that a Portuguese map must be based on Portuguese surveys might be an application of the provenance paradigm in our own time. Contrast the excellence of the Greenland outline with the relative poverty of likeness in the Newfoundland survey. Nevertheless, the islands along the west coast of Greenland are erroneous, belonging to the far north but not the southern west coast. This is not just an error but a degeneracy. Inaccuracies are characteristic of fresh explorations, but degeneracies are not. Perhaps some prototype for this map came down through João Vaz Corte Real.

One of the most striking aspects of the Cantino Map is its contrast with

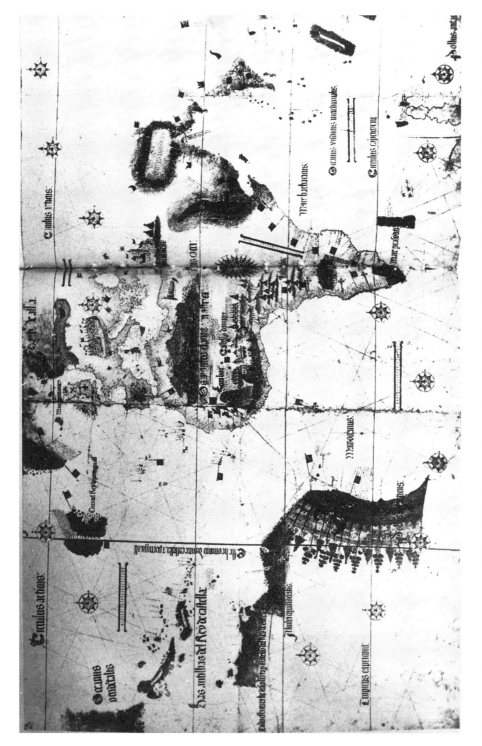

Figure 51. Anonymous map called the "Cantino map," 1502. Location: Modena, Biblioteca Nazionale Estense. Source: Cortesão, *Cartografia*, vol. 2, estampa 2.

the La Cosa map's treatment of northern America, for which it shows nothing but open water. Some writers have suggested that the mapmaker purposely moved his *Terra del Rey de Portuguall* away from the coastline. He may have wanted to bring it to the Portuguese side of the Tordesillas Treaty line, the heavy vertical meridian. (This had been agreed upon between Spain and Portugal, after much papal intrigue, to divide new discoveries between them.[1]) This accusation is unjust, however, for his east-west placement of Newfoundland is exactly correct relative to the longitude of the easternmost end of the Indies. This compass map uses no particular projection theory, but it does seem to have a scale distortion in the New World area. The mapmaker allotted the same distance from the Tropic of Cancer to the Arctic Circle as from the treaty line to the newly discovered Florida area. The latter distance should have been about half the former. If the cartographer had wanted to cheat in favor of the Portuguese, he would have made this distortion in the opposite direction. The plain fact seems to be that he had no reason to believe that there was mainland in the northern part of the Oceanus Occidentalis. The Corte Reals had expressed the possibility that their discoveries might be connected to the Spanish lands, but they had not sent back any surveys to that effect. He had no reason to place land there unless he had access to the hypothetical La Cosa prototype and interpreted it in the same way La Cosa did. Meanwhile, it seems inevitable that he would have been aware of the Cabot explorations. It seems almost inevitable that by 1502 he would have seen one of the apparently several copies which had been made of Cabot's planning map. The conclusion must be that he was unwilling to undertake the rationalization that, hypothetically, went into the La Cosa map.[2] This problem will be resolved in the discussion of the next map.

1506: Giovanni Contarini. Contarini was a Venetian mapmaker about whom nothing is known except what appears on this map (Figure 52), which bears an inscription reading: "The geography of Ptolemy to 180 degrees with the addition of the other hemisphere in the same order also on a plane of 180 degrees, and if by folding together the two sets of degrees you form them into a circle you will perceive the whole spherical world combined into 360 degrees. Made known by the industry of Giovanni Matteo Contarini and by the art and ingenuity of Francesco Roselli, of Florence, in 1506."[3]

The map shows the antipodal continent soon to receive the name America and also, in the north, the eastward projection similar to the one on La Cosa's map, that we surmised was based on a Cabot prototype. However, there is no suggestion of a geographical proximity of this projection to the

Figure 52. Giovanni Contarini's planisphere, 1506. Location: London, British Library Map Division, C.2cc.4. Source: Fite and Freeman, *Old Maps*, p. 18. Courtesy of Map Division, The New York Public Library, Astor, Lenox and Tilden Foundations.

West Indies. It is shown as a part of Asia identified as Marco Polo's province of Tangut. The map does, however, show a striking proximity of this projection to northern Scandinavia, and the sea between is almost bridged by chains of islands. Contarini made no attempt to associate this land with English discoveries, and a scroll beneath it attributes it to Portuguese sailors. This has frequently been assumed to refer to the Corte Reals. But there is no resemblance whatsoever between the land shown here and the survey of Newfoundland on the Cantino map and later maps. The Portuguese referred to in this scroll could just as well be those in the Müntzer letter of 1493, who presumably accompanied Pining and Pothurst. In fact, the same piece of land reported by La Cosa and Contarini was in one case attributed to the English (by La Cosa) with no mention of the Portuguese and the other time attributed to the Portuguese (by Contarini) with no mention of the English. Such a discrepancy seems uncharacteristic of any actual geographical discovery, but it may be characteristic of a paper map claiming to show a reality yet to be (re-)discovered. Such a map as this, preexisting, would explain nicely the theory of Cabot, who expressed in his patent of 1496 that he did not need to enter the southern seas in order to reach the lands of Marco Polo.

Let us look closely at the geographical qualifications of this unknown piece of land as drawn by Contarini and compare them with La Cosa's map (Figure 50). At the southeastern extremity of this projection there is a significant peninsula. It appears at the same place in La Cosa, although there it is less protrusive. Between this peninsula and La Cosa's map's northern edge, where his map runs off the page (perhaps from trimming),[4] La Cosa shows a single promontory flanked by two bays. An analogous promontory, rounded and less defined by invaginated bays, is present in Contarini's rendition. Contarini then shows an ocean to the north of his representation of the projection. The immediate southern coastlines on both maps are similar and have few major features.

It may be that La Cosa and Contarini were dealing with the same piece of land, but there are enough dissimilarities to rule out their working from identical prototypes or, alternatively, to show that their prototype was in a form that allowed room for individual interpretation. In addition to their dissimilarities of emphasis on coastline features, they differ totally on their emphasis of interior features. The only internal features Contarini shows are a few mountains and a river flowing through his peninsula. La Cosa shows none of these but dots the interior of his land with myriad lakes. Perhaps the most striking dissimilarity is in the overall orientation of the land. Contarini's east-facing promontories form a line oriented rather northeast-southwest, but La Cosa's are nearly north-south. Contarini's northern coastline follows so closely, almost stylistically, along a circle of latitude that one suspects that it might have been surmised. Perhaps the whole question of orientation was ambiguous in the prototype. Could it be that a different orientation from either La Cosa or Contarini would lead to a valid identification of the land?

If Contarini's land is detached from its base in Asia and rotated 90° counterclockwise, the upper coastline becomes a western coastline oriented north-south and the eastern promontories become northern promontories. There is exactly one plausible subject in the New World that fits these conditions, exactly one piece of land which has a *western* coastline and *northern* promontories. That is the Quebec peninsula.

Rotation of a map to secure identification is one of the methods most frowned upon by the historical cartography establishment. We will treat this identification as highly tentative, and our analysis will become even more speculative for the present. It is immediately apparent that the general shape of Contarini's and La Cosa's land bears no resemblance to that of the Quebec peninsula. The characteristic wedge shape is missing, regardless of the orientation. Nevertheless, let us examine some of the individual features. Contarini's northeasternmost promontory compares very well with the northwestern promontory of the Quebec peninsula in the Wolstenholme area (see front map). His southeasternmost peninsula, with its dual rivers, compares slightly less well with Quebec's Cape Chidley promontory east of Ungava Bay, with its Korok and George Rivers. At this point, La Cosa's version of the peninsula seems to be a better representation of the rather sharp Cape Chidley promontory. Between the southeastern and northeastern corners of their maps, both La Cosa and Contarini show a central promontory. This corresponds to the Cape Hope's Advance promontory on the west side of Ungava Bay. Contarini's northern shoreline contains two less-pronounced swollen protrusions, and these would correspond to the Portland promontory and Cape Jones promontory along Hudson Bay and James Bay.

Turning to interior features, note the stylized lakes of La Cosa's map. The interior of the Quebec peninsula is rather flat, and the water table is quite near the surface. The result is myriad lakes dotting the entire interior, which could only be depicted stylistically. On Contarini's interior, the mountains could represent the coastal range along the northeastern part of the Quebec peninsula, the Torngat range.

Its wedge shape is imparted to the Quebec peninsula by the eastern Labrador coast, which for some reason seems to have been entirely omitted by La Cosa and Contarini. Perhaps, if our speculations are correct so far, it was lost in a process involving regional maps, which are also subject to improper rotation.

The (unrotated) southern coastline of both La Cosa's and Contarini's New World Maps is a good representation of the true southern coastline of the wedge, the northern coast of the gulf and estuary of St. Lawrence. This is a relatively featureless coastline, except for rivers, and both maps show it so.

Down along the east coast of "Asia," Contarini shows the bulbous knob which had appeared on many maps since Bianco. This is the feature whose identification I have been postponing until a more refined rendering of it appeared. Such a rendering of it appears in La Cosa's map. There he replaces the rough bulbous outline with a gently hooked prominence sheltering two islands in its gulf. Except for the number of islands, this is a good representation of the Gulf of St. Lawrence. The gently arcing prominence conforms nicely to the gently arcing coastline of Nova Scotia, New Brunswick, and the Gaspé area. At the top of the arc one sees a river analogous to the St. Lawrence, emptying into the gulf from an inland lake. Contarini shows a river in the same relative position on his map, although it does not issue from a lake. Contarini shows a multitude of islands in the gulf compared to La Cosa's two, and this is correct for the Gulf of St. Lawrence. However, La Cosa's larger island might represent Newfoundland.

As was noted previously, La Cosa would by no means have been the first to map this area. The bulbous protrusion of "Asia" in Contarini's style had appeared at least since Bianco. Perhaps the bulbous version arose from attempts to represent Nova Scotia, which is indeed set away from the mainland by the Bay of Fundy. La Cosa seems to have had access to a different source of information, one which presented more precise detail of the entire St. Lawrence area. We may examine the Zeno map of Estotiland (Figure 67) and the country he calls Drogeo to its south. This Drogeo is in the correct position to represent the southern shore of the Gulf of St. Lawrence, and it is correctly studded with islands. The largest of these small islands has a discernibly crescent-shaped north coast, which is characteristic of Prince Edward Island. While the Zeno map cannot necessarily yet be considered as trustworthy, the Contarini map can.

1507: Martin Waldseemüller. The highly productive German cartographer Martin Waldseemüller worked around Strassbur. In 1507 he produced a large wall map about eight feet long on which the smaller hemispherical views were positioned above the north polar "glacial sea" (Figure 53). The Old World hemisphere is depicted as the geographical province of Claudius

Ptolemy, and the New World as that of Americus Vespuccius. Thus first emerged cartographically the idea that the new lands in the west were *not* Asia or the antipodes but a completely new world opposite to Orbis Terrarum. This idea was not immediately accepted by all cosmographers. The depiction of this new continent (not visible here), on Waldseemüller's world map is the first known to give it the name *America*. The cartographic influence this widely copied map enjoyed evidently caused the name to stick. Some question about this has been cast by observations that the form and spacing of these letters is different from others on the map and that they overlie meridian and parallel lines.[5] Nevertheless, an accompanying set of globe gores also carries the "America" inscription, and an accompanying text explicitly recommends the name.

More in the line of study here are two inscriptions along the Arctic coast of this "glacial sea," which at first thought would seem to refer to the Lapps. Far in the east is the notation "Balor Region. The inhabitants of this region live in the mountains, are rustic people, lack wine and self-sown wheat, use the flesh of deer and ride domesticated deer." Farther west, on a finger of land near what presumably represents the White Sea, is inscribed, "The inhabitants hereabouts have no leader, ever since the past being in beastiality. Valorous people are conjoined with great hazards. They lack not only wine and self-sown wheat but also rich kinds of meat."[6]

The concern in these inscriptions with "wine and self-sown wheat" immediately brings to mind a people other than the Laplanders. Namely it suggests the Vinlanders as described by Adam of Bremen, who in fact located them along the north coast of Asia. Note the suggestion in these inscriptions that the people farther east were more advanced than those to the west, nearer European civilization. This is quite the reverse of the situation of the Lapps and people beyond them. However, if one interprets these inscriptions as relevant to the Norsemen on the Arctic coast of the still unrecognized North America, then the sequence is correct. In the east are the American Norsemen, who may have tried to domesticate caribou. To their west are the Thule Eskimos, who were indeed anarchical and preferred sea mammals as food over land mammals. That a cartographer at the beginning of the sixteenth century would have such interests suggests a closeness to Norse sources.

1507–1508: Johann Ruysch. A Dutch painter who was born in Utrecht, Ruysch became a Benedictine monk. He was appointed to a station in the papal palace shortly before his already discussed map (page 000) appeared in the Rome edition of Ptolemy in latter 1507 and, with accumulating modi-

Figure 53. Old World hemisphere, by Martin Waldseemüller, 1507. (Detail from larger wall map.) Location: Würtemberg, Library of the Princes of Waldburg zu Wolfegg-Waldsee. Source: Kamal, *Monumenta Cartographia,* vol. 5, fasc. 1, fol. 1513.

fications, throughout 1508 (Figure 54). Concerning his activities as a map-maker contemporary references are mute, except a short statement by the editor, Marcus Beneventanus, in his preface to the Rome Ptolemy: "Johann Ruysch Germanus, in my judgment surely the most skillful of geographers, and in painting maps the most painstaking, whose support we have used in this little lucubration, claims personally to have sailed from the southern part

of England to the latitude of 53° north, and in this parallel to have sailed to the shores of the East through the hidden realm of the Night, as well as to have surveyed more islands." The phrase "the hidden realm of the Night" refers to the back hemisphere of the Earth. This evidently still had a certain amount of mystical fascination for readers as a way of reaching the Orient.

Ruysch's asserted oriental shores are almost identical to those in Contarini, and he uses nearly the same folding conical projection as Contarini did. His rendition of the Quebec peninsula analogue also seems to be related to the one in Contarini. Nevertheless, whether or not Contarini's map served

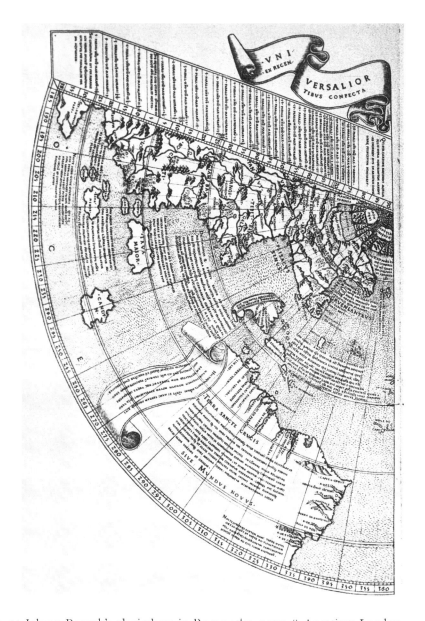

Figure 54. Johann Ruysch's planisphere in Rome atlas, 1507–8. Location: London, British Library Map Division, C.1.d.6. Source: Nordenskiöld, *Facsimile Atlas,* pl. 32. Courtesy of Map Division, The New York Public Library, Astor, Lenox and Tilden Foundations.

to inform Ruysch, the latter has impressed his own interpretation on the data. He has reduced Contarini's east-west extension of the land and made room for Newfoundland, imported from the Cantino map, at the proper distance from England. This *Terra Nova* he has attached to Contarini's wide peninsula, perhaps reflecting the old idea of the Bristol merchants that the

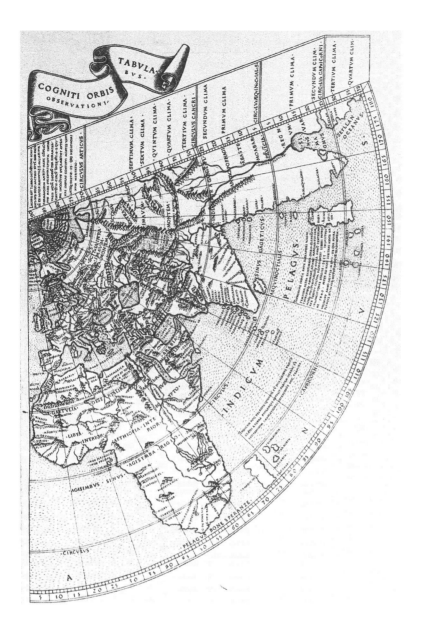

Isle of Brasil was connected to the mainland. The word *Baccalauras* was also to become associated with the Newfoundland–Grand Banks area, being derived from the Basque word for "codfish." If Ruysch did actually make the voyage to which Beneventanus gives him claim, then it is difficult to see where he did his surveying. The fifty-third parallel on Ruysch's map just skirts the northern end of Newfoundland, and all the coastlines in that area are evidently copied from previous maps.

Farther north, however, Ruysch shows many unique coastlines, and surrounding the pole he shows the Septentrional Isles of the *Inventio Fortunatae.*

He certainly shows "more islands" than the standard map of the time showed. According to Cnoyen they were surveyed in a systematic procedure and were so presented in the *Inventio.* In a marginal inscription, Ruysch says: "It is to read in the book *De Inventione Fortunati* that under the Arctic pole there is a high rock of magnetic stone 33 German miles in circumference. This is surrounded by the Indrawing Sea, flowing as if in a vessel that lets water down a hole. There are four surrounding islands, of which two are inhabited. Encircling these islands, however, are confining mountains, twenty-four days journey across, which deny human habitation." Ruysch may have accompanied some voyage to the new found land, but he certainly did not go to the Arctic Archipelago. Beneventanus, however, may have gotten that impression from the authoritativeness with which Ruysch spoke on the subject and the uniqueness of the subject. Ruysch would not have been the first to receive undue credit for the *Inventio* voyage.

Aside from the Septentrional Isles, Ruysch shows strongly defined features on the north coast of the continent. These features seem rather stylized, nevertheless, and escape identification. His version of Greenland is melded into the Quebec peninsula analogue, so it becomes the ultimate peninsula of Asia.

In the south, Ruysch depicts recent New World discoveries, but he shows no evidence of any belief that this land was connected to the lands in the north. The map of 1507 by Martin Waldseemüller also leaves the northern ocean completely open, even omitting the Quebec peninsula analogue. The same situation is present in the Cantino map. The following conclusion seems tenable: La Cosa's implication of a continuous coastline from north to south (page 221) was spurious and accidental and did not reflect any well-founded belief of the time. The consensus of the time still seems to have been that there were many westward passages to Asia available. There was not yet a requirement for any secret passages around the northern part of an intervening continent.[7]

1509: Sebastian Cabot Voyage. Sebastian Cabot, the oldest son of John Cabot, had a long and active life. Its adult portion spanned the period from his father's first discovery of American mainland until 1557, by which time the American seaboard was completely mapped. Sebastian was intimately associated with this gradually accumulating body of new knowledge. He spent the major part of his career, from 1512 to 1548, in high political posts in Spain charged with maintaining royal geographical knowledge. Both before and after this he was active in English expeditions. Perhaps the sensitive nature of his position was augmented by a natural reticence, because he made few statements for publication. Nevertheless, contemporary historian recognized the great value of his knowledge. After Cabot's death Richard Hakluyt announced, "Shortly, god willing, shall come out in print, all his own mappes and discourses, drawn and written by himselfe, which are in the custodie of the worshipfull master William Worthington, one of her Majestie's pensioners, who (because so worthie monuments shoulde not be buried in perpetuall oblivion) is very willing to suffer them to be overseene and published in as good order as may be." The irony of the situation is that for some unknown reason this invaluable collection was lost before the promised publication and never heard of again. As a result, our knowledge of Sebastian's activities must be derived from sparse accounts by other writers, who are often contradictory and who wrote much after the activities they described. The situation is further complicated by an apparent streak of vanity in Sebastian; he refrained from denying a common misunderstanding that gave him credit for his father's activities.

What can be discerned with reasonable certainty starts during his pre-Spanish residence in England. There he served King Henry VII sufficiently well to merit the award of a ten-pound annuity starting in 1505. A record recently discovered by Alwyn Ruddock makes it clear that the award was for explorations in Newfoundland in 1504. Furthermore, there is record that Cabot was away on a voyage when Henry VII died, in 1509. One of the more detailed accounts of this voyage (Gomara's) was written in 1552:

> And he promised King Henry to go by the north to Cathay, and to bring thence spices in a shorter time than the Portuguese did by the south. He went also to learn what sort of land the Indies were to inhabit [colonize]. He took three hundred men and went by way of Iceland to the cape of Labrador [Greenland]* until he reached fifty-eight degrees, although he himself says much more,† relating that there were in the month of July such cold and so many pieces of ice that he dared not go farther; and that the days were very long and almost without night, and the nights very bright. It is a fact that at sixty degrees the days are eighteen hours long.

The idea of going by the north to Cathay was highly practical in 1509. Except for La Cosa's, all the maps of the time showed vast open waters there. It

---

*Through a complicated historical process, the name *Labrador* was first applied to Greenland and only later became displaced to its present meaning. When Gomara was writing, this would still have referred to the 1509 meaning of *Labrador,* namely, "Greenland."

†Depending on which writer is claiming to quote Cabot, the latitude reached is cited as various degrees in the sixties.

Figure 55. Unidentified chart known as "Kunstmann III," ca. (?)1509. Location: Munich, Bayerische Armee-Bibliothek, no. 31. Source: Björnbo and Petersen, *Anecdota,* pl. 4. Courtesy of Map Division, The New York Public Library, Astor, Lenox and Tilden Foundations.

could easily be demonstrated with a globe that the great circle distance from England to China via the northwest was far less than any western passage would be.

Sebastian Cabot's voyage was turned back, not by any encounter with obstructing land, but by the ice. Quotations by a later writer (Richard Willes) describing a now-lost later map of Cabot's make it clear that they had sailed right up into Hudson Strait and viewed Hudson Bay (see front map): "as in his owne discourse of navigation you may reade in his carde drawen with his owne hands, the mouth of the north-western streict lieth neare the 318 Meridian, betwixt 61 and 64 degrees in elevation, continuing the same breadth about 10 degrees west, where it openeth southerly more and more."

There is an extremely interesting aspect of Cabot's presumed penetration of Hudson Strait in 1509. At that time, even though the maps showed much

open water in the northwest, his plotting a course by any of them would surely have led him to dry land. Furthermore, none of the reports suggests that he had to search for the strait but that he went there rather directly. The probability of this happening by accident would seem quite small. There is another unusual aspect to the voyage. Instead of starting his Atlantic crossing from Bristol and following the usual established route, or even sailing northwest from England, he rather took a route via Iceland and Greenland, and thence westward. Explorers are not known to have sailed such a course since the old Norse days, with one known exception.* That was Johannes Scolvus in 1476 (page 196). A further interesting coincidence is that references to Scolvus are nonexistent before John Cabot's time, with most of the known references succeeding Sebastian Cabot's later attempts to popularize the passage. The circumstantial evidence suggests that Sebastian Cabot was following old Norse sailing directions, and in this event, loss of his collected papers is doubly regrettable.[8]

---

*Seaver would have us believe that there were many other exceptions, but her arguments are highly speculative.

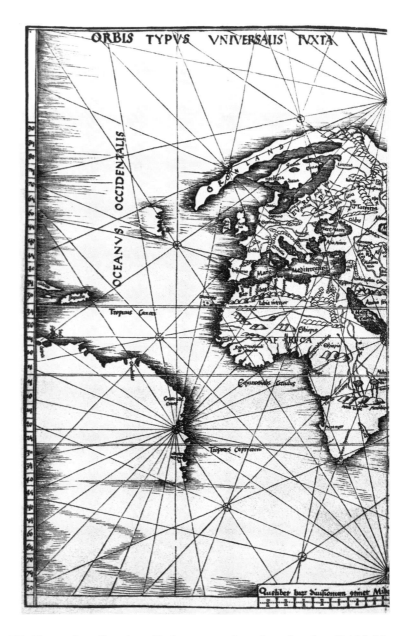

Figure 56. World map from Strassburg Ptolemy, 1513. Locations: London, British Library Map Division, C.1.d.8; Washington, D.C., Library of Congress Map Division, no. 359. Source: Nordenskiöld, *Facsimile Atlas,* pl. 35. Courtesy of Map Division, The New York Public Library, Astor, Lenox and Tilden Foundations.

1509(?): Kunstmann III. Sometime between 1503 and 1511, an unknown Portuguese cartographer[*] drew the map shown in Figure 55, now known by

---

[*]Previously erroneously thought to be Pilestrina.

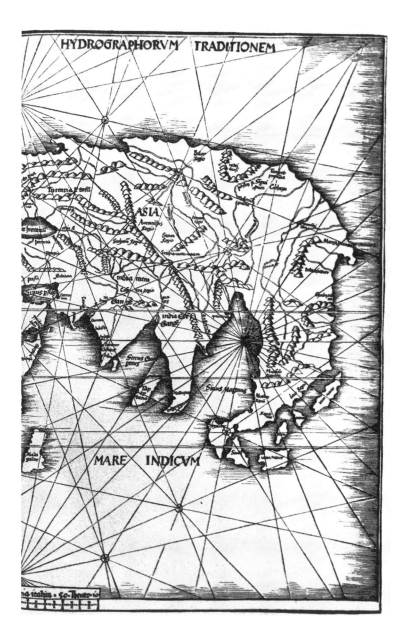

a photographic plate number assigned by F. Kunstmann, one of the first modern scholars to study it. The map is unsigned and undated.[9]

   This map's depiction of the western part of the North Atlantic is a very good contemporary survey of the Gulf of St. Lawrence and adjoining coasts, even showing signs of hearsay knowledge of the Great Lakes. This survey is difficult to attribute to the Corte Reals, but on the other hand all subsequent Portuguese in the area were simple fishermen, who neither wanted to nor could have gone on a mapping expedition along the coast. However, Williamson has shown that there may have been Anglo-French cooperation during that era, and it is known that several English exploratory voyages were sent out at that time. The most likely one to have produced this mapping would seem to be that of Sebastian Cabot. After being stopped by the ice, he could have rescued the voyage from uselessness by surveying southward along the coast, according to some reports below 40° latitude. He is likely to have produced an accurate map as a result.

North of this land is what appears to be a representation of Greenland, which might be a combination of actual contemporary survey and Norse prototype or hearsay. This identification also fits in with the details of Sebastian Cabot's voyage, which actually went to Greenland. In the dark regions north of the Arctic Circle the map shows a highly intriguing feature; circling the pole is a stylized ring of mountains, which presumably was taken from the *Inventio Fortunatae*. The land on which these mountains lie, however, has no precedent. While the left part seems to bear some resemblance to Contarini's land, this might be accidental, because of its position. The cartographer's primary goal was to show a strait through the ring of mountains between the two continental land masses. Now, while the North Atlantic has no such strait, Clavus showed nearly a century before that the Norsemen had information about such a strait elsewhere—the Bering Strait between Alaska and Siberia. Under the one-ocean paradigm's perception of Alaska and the current perception of eastern Asia, the North Atlantic is exactly where that strait should be placed.

This map has other features that seem to tie it to Clavus. First, to the east of the strait where the continental coastline begins to come southward, there appears a string of toponyms that are in fact Clavus's folk-song names from his second map. To the west, on the east coast of what we surmise to be Greenland, there are exactly three named capes (there may be traces of a fourth, erased one). These names are not familiar in Greenland and seem contemporary with the map. However, recall that on Clavus's first map his Greenland had exactly three quantitatively specified coastal points. It is possible that we are here seeing a map derived from the postulated lost third map of Clavus, with its "great Arctic strait and the land beyond." This will be addressed in the next chapter.

1513: Strassburg Ptolemy. The so-called Argentinae edition of Ptolemy published at Strassburg in 1513 contains two world maps. One is of the classical Ptolemaic style and the other (Figure 56) is of the modern mariners map tradition. The latter suggests a continuing influence of the misunderstanding of

Alaskan land as belonging in Scandinavia. The land labeled Greenland is of the A-type (see page 199) and seems to be a composite of accounts about the traditional Greenland and the Aleutian Peninsula. This map is believed to have been made by Martin Waldseemüller. If so, it represents a change of mind regarding his depiction of Greenland, which was of the B-type in his 1507 map, reflecting the Quebec peninsula. A 1516 map of Waldseemüller's, not shown here, is of the A-type.[10]

SUMMARY

A rather broad variety of information on familiar Norse areas in the eastern and central Arctic persisted during this decade of South America's cartographic acceptance. There was further evidence of a continuity between a recognizable Nova Scotia and the bulbous feature on pre–Corte Real maps of east Asia, suggesting Norse-transmitted knowledge of areas at least as far south as Nova Scotia. Since the Eskimos never went there, this supports the thesis that Norsemen or their European guests explored down the eastern seaboard themselves or at least went far enough to meet Indians with that knowledge.

CHAPTER 15

AN OLD CONTINENT EMERGES

AS WE HAVE SEEN, NORTH AMERICAN LAND HAD, over a period of some centuries, appeared on European maps of the world. As explicit information on this yet to be recognized continent accumulated, geographers no longer had the option of dealing with Norse information by simply tucking it into the Far West or Far East. Increasingly, the only *terra incognita* left on the map in which to place Norse information was the Arctic region, and that is just where much of it belonged. This situation seems to have led to a prolongation of the misunderstanding of Arctic America as Arctic Eurasia even while a modern picture of the American eastern seaboard was developing.

1515 and 1520: Early Johann Schöner. In later years a renowned professor of mathematics at the *Gymnasium* in Nuremberg, Johann Schöner in his early years published numerous mathematical and astronomical treatises and corresponded with Copernicus. In 1515, holding the honorary title *praetor* at Bamberg, he published the first known of a series of globes (Figure 57). It had an accompanying descriptive text, *Luculentissima Descriptio.* In this text, Schöner makes it clear that his source for the Scandinavian regions was Claudius Clavus. He makes many statements that seem to be exact quotations from the Vienna text of Clavus. On the other hand, his statements are in a completely different sequence from the Vienna text. Also, in many cases where the informational content is the same, the wording has been changed. While this would normally be ascribed to Schöner's editing, there are some points that lead me to examine an alternate possibility. That possibility is that Clavus made a later edition, from which Schöner worked.

The Vienna text ascribed a longitude of 41° to Ynesegh, the market town which Clavus described as lying right on the Arctic Circle (page 45). Schöner

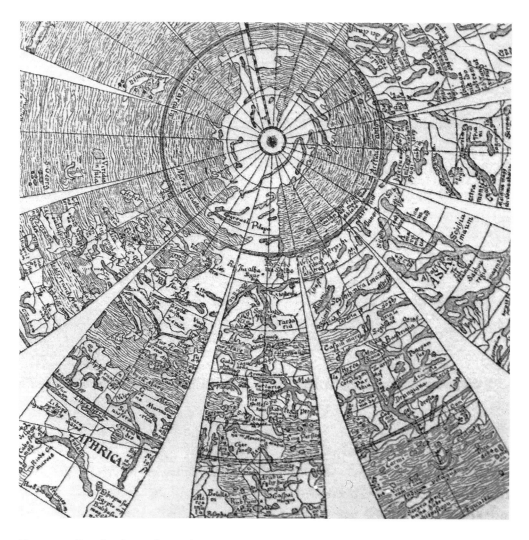

Figure 57. Detail of gored Northern Hemisphere map by Johann Schöner, 1515. Locations: Weimar; Frankfurt am Main. Source: (After Jomard) Nordenskiöld, *Facsimile Atlas,* p. 78, fig. 46. Courtesy of Map Division, The New York Public Library, Astor, Lenox and Tilden Foundations.

quotes its longitude as 38°. This kind of change cannot be ascribed to editorial style. Neither can it be ascribed to Schöner's informed correction. At that coordinate point on his globe there is nothing but open water. Evidently he faithfully copied it from his textual source without checking its plausibility on his globe. Similarly, the Vienna text describes a promontory called Nadhegrun at 30°35′ longitude, which Schöner's text changes to a precise extension from 26° to 31°. These changes are not apparent in another contemporary writer, Friedlieb, who evidently worked from the Vienna text itself, the only discrepancies being obvious copying errors.

In addition to the changes, Schöner's text omits many lesser Scandinavian toponyms that the Vienna text lists. Schöner himself would, of course, have been in a much poorer position to make such decisions with authority than would Clavus, who may have been recanting his made-up names. Besides all this, there are further reasons to believe that Clavus made another version after writing the Vienna text. The early maps of Donnus Nicholaus Germanus, known to scholars as the "A-type," are exactly derived from the Vienna text of Clavus. Later maps, known as the "B-type," are also derived from Clavus but differ from the Vienna text in two major areas, discussed earlier under Nicholaus (page 199). Schöner's map is of the B-type. The simplest assumption is that B-type maps were taken from a later work of Clavus that has since been lost. The Schöner text could well have been taken from such a work. It includes passages that are not part of the Vienna text but are included in the texts accompanying other B-type maps.

This observation now brings us to consideration of Schöner's north polar area on his printed globe of 1515. While the B-type maps all have the characteristic wedge-shaped "Engroneland," no other map has shown continental land right under the pole star. Schöner himself makes no special mention of it in his text, evidently accepting it as covered in Clavus's description of Engroneland: "It stretches down from land on the north which is inaccessible or unknown on account of ice."

This is perhaps the strongest reason of all for believing that Schöner worked from a third text (and map) of Clavus. If he had any other source for his polar land he would have been bound to mention it at this point. Instead, his text turns directly to another subject. The Vienna text, however, after making the above statement, continues: "Nevertheless, as you may see, the pagan Karelians daily come to Greenland in a great flock, and that without doubt from the other side of the North Pole. Therefore the ocean does not wash the limit of the continent under the pole itself as all ancient authors have asserted; and therefore the noble English knight, John Mandeville, did not lie when he said that he had sailed from the Indian Seres to an island in Norway."

The polar continent on Schöner's globe seems perfect to replace this passage in the Vienna text. If Clavus had had such a map, he would have done well to let it speak for itself and drop any mention of the pagan Karelians and the more notorious than noble John Mandeville. A possible prototype for the map may be discerned by facing it from its base in "Engroneland" and "Pilapa[land]." Above this one sees a circumpolar continent bifurcated by a deep bay opposite the North Pole. This bay has some of the characteristics of a greatly narrowed Hudson Bay, with a suggestion of James Bay at the bottom. To its right is a sector of land that is set off by another lesser inlet about 70° clockwise from the Hudson Bay analogue. This could represent a doubled-scale analogue of the wedge-shaped Quebec peninsula. Left of Hudson Bay would be Melville Peninsula jutting upward and the coastline then continuing to the left. In the upper left corner jutting out into the sea is a configuration which, at this scale, is impossible to identify with anything real. But it could be an agglomeration of the entire Arctic Archipelago into a single land mass, connected to the continent by the Boothia Peninsula. The modern scholar Dana Durand made just such an agglomeration in a Vienna-Klosterneuburg map (page 168), and some versions of the United Nations emblem, a world polar map, contain an agglomeration of this archipelago.

These interpretations are corroborated when we realize that the left side of the polar continent does not actually proceed southward in this rendition. It merely follows a circle of constant latitude to the westward. The displacement of the land relative to the true pole has caused the map of this North American Arctic coastline to be convex rather than concave. After overcoming this disconcerting effect, we come to the interesting realization that this kind of distortion could only have arisen from data containing latitude observations, and it therefore implies European presence in the area. The two peninsulas along the left side now correspond to Behaim's two peninsulas, which have been identified as Adelaide Peninsula and Kent Peninsula (page 59). We may be seeing the *Inventio Fortunatae* speaking again, this time filtered through Clavus.

In 1520 Schöner produced a hand-painted globe (Figure 58) which, except for small details, was an exact copy of the 1515 printed globe. The changed details of interest to us are in the polar regions and in the easternmost promontory of Asia. The 1515 version represents this promontory as a simple, undetailed peninsula, but the 1520 version adds details like those we discussed in comparing La Cosa and Contarini (page 238). Schöner's *Quian* river would correspond to the St. Lawrence, the largest river in the neighborhood. On the south shore of its outlet a peninsula points northeastward, corresponding to the Gaspé Peninsula. The bulbous land below that would correspond to New Brunswick and Nova Scotia. Continuing the analogy, at the lower end of the bay below putative Nova Scotia is a peninsula pointing toward Nova Scotia that would correspond to Cape Cod. The north polar coastlines are essentially identical to those of the 1515 globe except for small changes in the outline of the agglomerated Arctic Archipelago. However, just along the coastline of the polar mainland is an inscription that was not pres-

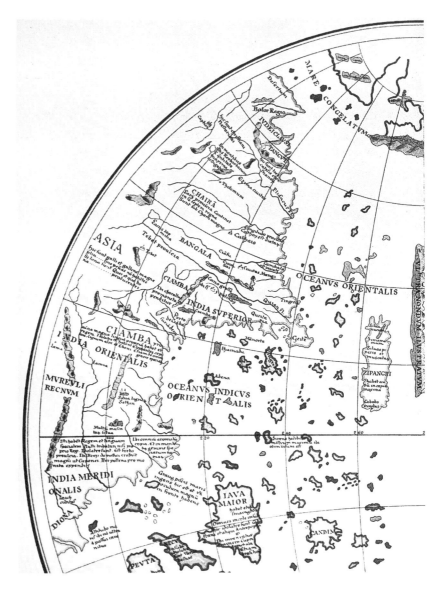

Figure 58. Tracing of Western Hemisphere from Johann Schöner's 1520 globe. Location of globe: Nuremberg, National Library. Source of tracing: Kretschmer, *Entdeckung,* pl. 13. Courtesy of Map Division, The New York Public Library, Astor, Lenox and Tilden Foundations.

ent in 1515: "This polar arc is said to be desert for 3000 or more German miles." The treeless tundra of the North American Arctic coastline in fact stretches just over 3000 English miles, from northern Quebec to Alaska.

The polar continent here reminds one of the mystifying comment by Toscanelli's friend Plethon (page 198) about a map (presumably by Clavus) showing "the great arctic strait opposite the Arctic continent and the land

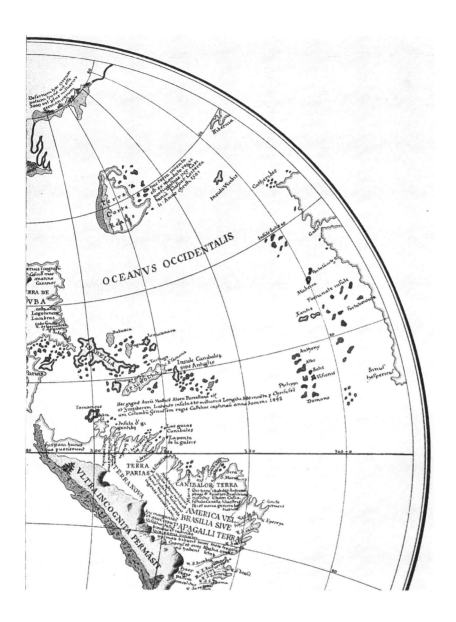

beyond." The first thought might be that the phrase refers to the Skagerrak and Kattegat between Denmark and Scandinavia, but this strait is hardly Arctic and no other writers, including Clavus, are known to have referred to it with such a phrase. Just such a phrase was used, however, by later writers who were searching for the "mythical Anian Strait" of the Northwest Passage. That strait did not receive its name (lifted from Marco Polo) until Jacopo Gastaldi used it in 1562. But it appeared cartographically at least as early as 1530 in an anonymous sketch map (Figure 59) and probably ca. 1509 in Kunstmann III (Figure 55). Note the similarity of the plunging coastal arc in both

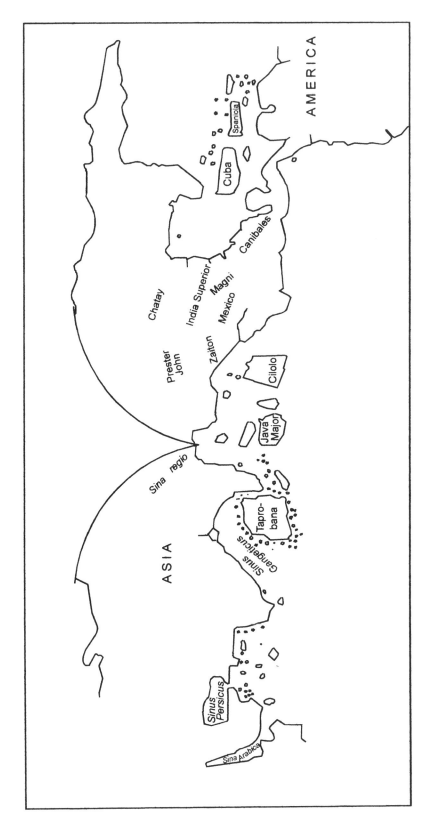

Figure 59. Sketch showing Great Arctic Strait, ca. 1530. Location: London, British Library, Sloane MS 117, fol. 3v–4. Source: G. Sykes, "The Mythical Straits of Anian," *American Geographical Society Bulletin* (1915), vol. 47, fig. 2. (Author's tracing.)

(rendered with a compass in the sketch map), suggesting a common, perhaps verbal, descriptive prototype. The coastlines of the Chukchi Sea (front map) do indeed have such a shape. Modern writers refer to the "mythical" Anian Strait, but earlier writers were absolutely sure of its existence. Clavus's first map has suggested that information on the Bering Strait was available, even if misinterpreted by him. In fact, when Clavus superseded his first map with his second, he would then have been confronted with the problem of reinterpreting or explaining away the prototype used for his first map. "If that first prototype was not Scandinavia, then what was it?" he must have asked himself. Perhaps the prototype from the first map, the Bering Strait, was somehow worked into the third map. Perhaps it was such a map that was known to Toscanelli and Plethon.[1]

1523–1536: Magellan and the Schöner-Finaeus Series. In 1523 Schöner was cultivating the favor of an official in the bishopric of Bamberg. He wrote the official a little letter summarizing the most up-to-date knowledge of geography. The new explorations generating this knowledge would bring about a thorough transition in cartography. There would soon be no place for maps based on misunderstood Norse-supplied prototypes, and today's view of the world would soon emerge. After describing the Spanish explorations from Columbus through the 1515 voyage of Cortés, the letter continues:

> For the other voyages of discovery, the admiral selected was Ferdinand Magellan, a Portuguese, well experienced in navigation, who set sail on 10th August, 1519, but met with an untimely end, as did his successor John Serrano; but Serrano's successor conducted his squadron through the most remote parts of the ocean, so that he sailed completely round the world in three years' time. What dangers and inconveniences he met with will be easier for you to imagine than for me to describe at length. Having sailed about in various directions, so as to leave no portion of the route unexplored, he returned to Spain, where he arrived on 6th December [error for September] A.D. 1522; accompanied by eighteen soldiers, and in one ship, the rest having been lost at sea. What wonderful adventures and what extraordinary men and animals they beheld, your Reverence will read at length in the epistle relating to the Molucca Islands, addressed by Maximillian Transylvanus to the Cardinal Archbishop of Salzburg.
>
> Being desirous to make some small addition to this wonderful survey of the earth, so that what appears very extraordinary to the reader may appear more likely, when thus illustrated, I have been at pains to construct this globe [accompanying the letter], having copied a very accurate one which an ingenious

Spaniard has sent to a person of distinction. I do not wish however to set aside the globe I constructed some time ago, as it fully showed all that had, at that time, been discovered; so that the former, as far as it goes, agrees with the latter.

This globe of 1523 has been lost, although several writers have erroneously ascribed its identity to various anonymous globes of that period. What aspect of Schöner's earlier globes would he have found lacking in light of Maximilian Transilvanus's description of Magellan's voyage?

> Not that it was impossible *prima facie* to sail from the West round the southern hemisphere to the East; but that it was uncertain whether ingenious Nature, all whose works are wisely conceived, had so arranged the sea and the land that it might be possible to arrive by this course to the Eastern Seas. For it had not been ascertained whether that extensive region, which is called Terra Firma, separated the Western Ocean from the Eastern; but it was plain that that continent extended in a southerly direction, and afterward inclined to the west. Moreover two regions had been discovered in the north, one called Baccalearum from a new kind of fish, the other called Florida; and if these were connected with Terra Firma, it would not be possible to pass from the Western Ocean to the Eastern; since, although much trouble had been taken to discover any strait which might exist connecting the two oceans, none had yet been found.

Schöner had already shown in his earlier globes a passage such as Magellan took around the south of America,[2] and he needed no change there. However, his depiction of the northern part of the New World showed any number of passages from the "Western Ocean" to the "Eastern Ocean," contradicting the experience cited by Transilvanus. In particular, Schöner's earlier contention that America was a fourth part of the world, insulated by the sea from the other three continents, was in doubt. Transilvanus increased this doubt with his description of a plan the voyagers adopted when they developed trouble off an island in the East Indies archipelago:

> Soon after our men had sailed from Thedori, the larger of the two ships sprang a leak, which let in so much water that they were obliged to return to Thedori. The Spaniards seeing that this defect could not be put right except with much labor and loss of time, agreed that the other ship should sail to the Cape of Cattigara, thence across the ocean as far as possible from the Indian coast, lest they should be seen by the Portuguese, until they came in sight of the southern point of Africa, beyond the Tropic of Capricorn, which the Portuguese call the Cape

of Good Hope, for thence the voyage to Spain would be easy. It was also arranged that, when the repairs of the other ship were completed, it should sail back through the Archipelago and the Vast Ocean to the coast of the continent which we have already mentioned, until they came to the Isthmus of Darien [Panama], where only a narrow neck of land divides the South Sea from the Western Sea, in which are the islands belonging to Spain.

This ship evidently had no hope whatever of making use of the passages Schöner's earlier globes showed in the northern part of the New World. Schöner would have done well to erase immediately any false hopes raised by keeping those passages. The vessel that attempted to return via America was never heard from again.

How does one go about erasing false passages? How should one fill them in, and where should one stop? Should one, in fact, stop anywhere? The maps of Contarini and Ruysch had already shown the Baccalearum region as a projection of eastern Asia. Schöner's geographical interests are sure to have brought him in contact with a copy of the 1508 Ptolemy in which Ruysch's map appeared. The most straightforward way of eliminating the passages would be to amalgamate all the lands known and attach them to the east coast of Asia. This is exactly what Schöner did in his next known globe, that of 1533 (Figure 60). He explains this action in its accompanying text as based on the report of Magellan's voyage. Henry Harrisse suggested that the incorporation of such a configuration was the essential change in the lost 1523 globe from its predecessors.[*]

If Harrisse was right, then the 1523 globe was the progenitor of a long list of globes and maps which are nearly identical to Schöner's subsequent 1533 globe (a few shown in Figures 61–64).[†] The two planar maps Figures 61 and 64 are works of the Parisian mathematician Orontius Finaeus, who was greatly interested in the development of precise projection systems. The authorship of the Paris gilt globes and the Stuttgart gores is unknown.[3] With this new configuration, the New World was closed not only to false passages to the Orient but to any further accretion of misunderstood Norse prototypes. The eastern seaboard from this time onward was mapped with newly acquired and constantly improving data. And therewith finally came the end of the one-ocean paradigm as a means of assimilating Norse information.

---

[*]The separate naming of the land mass newly formed by this amalgamation did not occur until Mercator's cordiform map of 1538, not shown here. In that map, Mercator was the first to use the terms *North America* and *South America*.

[†]The earliest surviving such map, not shown here, was by Franciscus Monachus in 1527.

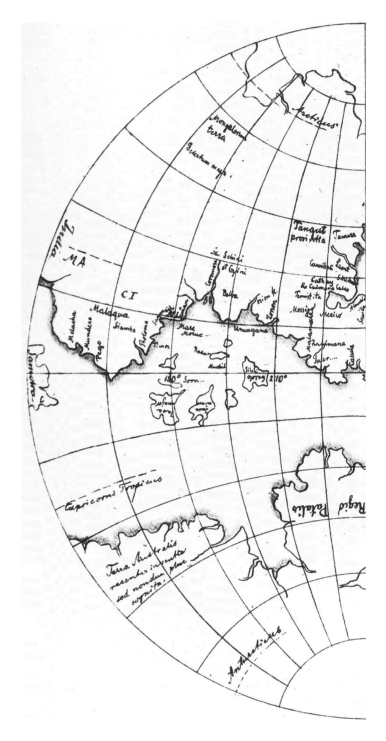

Figure 60. Western Hemisphere, traced from Johann Schöner's 1533 globe. Location of globe: Weimar, Military Library. Source of tracing: Harrisse, *Discovery*, pl. 17, opp.

75

60

Bachalaos

45

30

Oceanus
Occidentalis
Cancri Tropicus

15

Anegada
El anguila

Guadelupo
Dominica
Martinino ad Mart
Ia cera
drali

Baluc
R. Strale
Rio de Spirita
Terra Florida

Anegada
El anguila

Spagnolla   S. Iuan
Cuba
Iucatan
Iamaica
Gunolla

R. J. Amero
Cabo de Higo
S. Pablo
Sinus
Trabe
Rimu
Darium
Panuco
Angla

40   2 70   3 00 0   330

Temiscanara

Bralio

America   Indicus Superior sit Asia
continual: para

Marra de gra
R. Ilual
Mon: Argro

150

Sinulus
inferior
Pcompom

Prasili: mons

R. H. tanle
Baraso
R. S. lucia
R. Ia lam

Mare Magelani :
sum

R. S. Selera
C. S. Maria

80 °

Prima
Con.

Sinus
Cristian

45 °

60

75 °

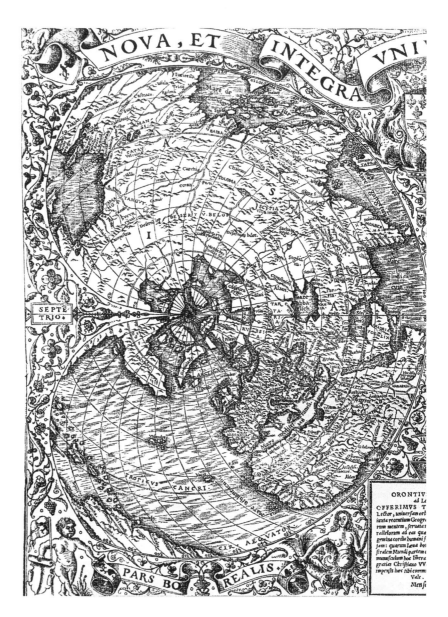

Figure 61. Double cordiform planisphere by Orontius Finaeus, 1531. Locations include: Ottawa, National Museum of Canada. Source: Nordenskiöld, *Facsimile Atlas,* pl. 41. Courtesy of Map Division, The New York Public Library, Astor, Lenox and Tilden Foundations.

The geographical depiction of the Arctic area in these maps and globes seems motivated by Ruysch's map of 1507–8. But here there is a much less stylistic interpretation of the source, suggesting perhaps more detailed source information. One might therefore hope for an identification of some of the naturalistic features of the Arctic coastline not possible in Ruysch's map: The

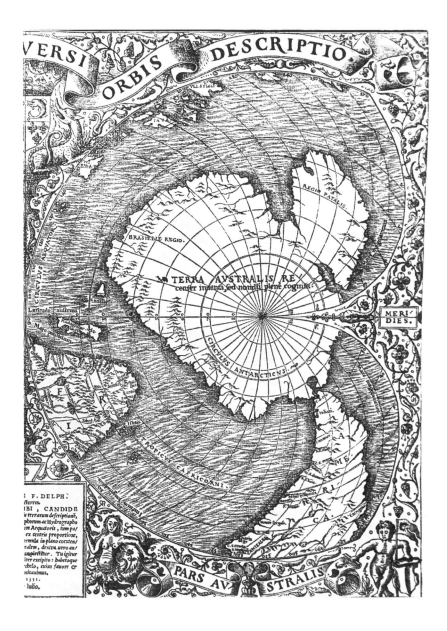

sharp north-pointing promontory adjacent to "Mare Tabin" (see Figure 63) seems a good likeness of the Cape Chidley promontory. The "Mare Tabin" is a good likeness of Ungava Bay. Cape Chidley at the end of the Quebec-Labrador peninsula is a good candidate for the ultimate cape of North America in the Arctic, as predicted by Pliny. The classical conception of Tabin peninsula did not provide any descriptive details, and the detailed agreement of this map with the Cape Chidley promontory and Ungava Bay makes the identification seem entirely reasonable.

While Schöner's globe of 1533 does express islands surrounding the pole,

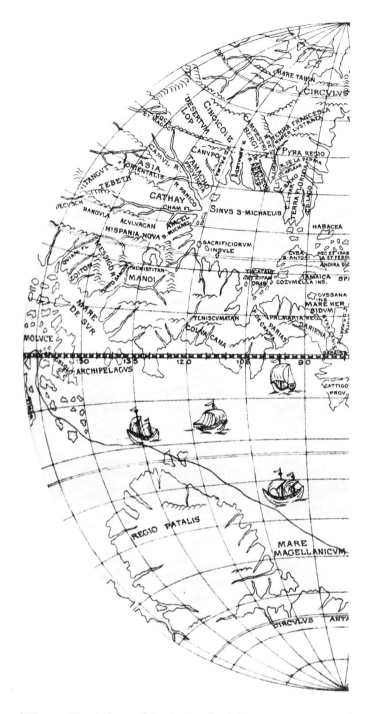

Figure 62. Tracing of Western Hemisphere of the Paris Gilt Globe, 1530s. Location of globe: Paris, Bibliothèque National de France Geography Department, no. 387. Source

of tracing: Harrisse, *Discovery,* pl. 21. opp. p. 562. Courtesy of Map Division, The New York Public Library, Astor, Lenox and Tilden Foundations.

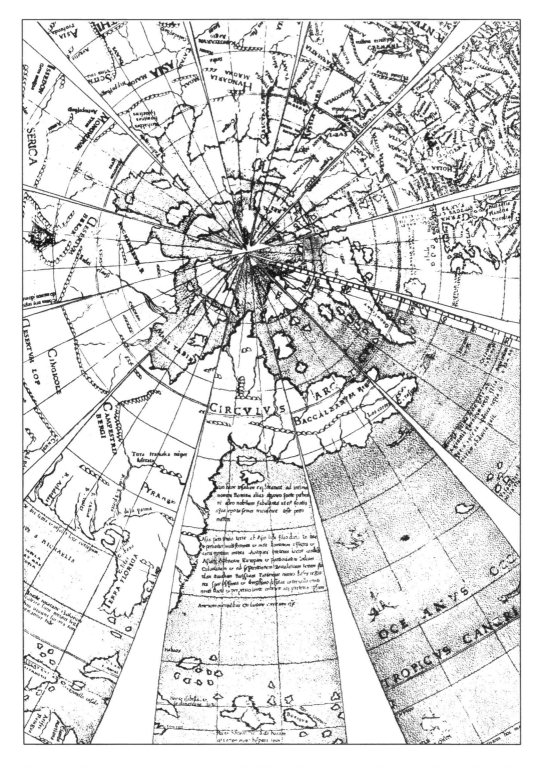

Figure 63. Stuttgart gores, 1530s, reassembled by author. Location: Stuttgart, Würtembergische Landesbibliothek, Nic. S. 78.56. Source: Wieder, *Monumenta Cartographica,* plates 1–3.

he also clung to the idea of continental land in the north. Perhaps the contemporary information on North America that had accumulated by 1533 blended in his mind with Norse information and from this melding he postulated a land bridge from northern Europe to northern America. But the coastlines appear to be based on data rather than conjecture. Several subsequent geographers repeated this bridge with or without interrupting straits or channels.

1532, 1539: Jacob Ziegler and Olaus Magnus. Jacob Ziegler was a Bavarian theologian, mathematician, and geographer. In 1532 he published his first edition of a geographical compendium; it included a description of the North, *Scondia,* and a map of Scandinavia (Figure 65). For the first time, a map represented the actual geographical reality of Scandinavia itself. It is obviously based on actual knowledge of Scandinavia without contamination by prototypes from diffuse northern sources in America. Of his specific sources of Scandinavian geographical information Ziegler says:

> Gothia, Suecia and Finland, to the north of which I have extended Laponia, the Greenland chersonesus and the Island of Tile, I have taken over from the reverend bishops Johannis Magnus of Upsala and Peter Aorosien of Gothland, personal friends in the city at that time and in the greatest confidence of social intercourse with me, and indeed the Upsalian in an account of Scandia which had previously been written, who had submitted it to our examination; I, who contemplating protecting the foundations of latitudes and longitudes from incursion, have converted into them from the system of azimuths and distances they were kept in, to the nearest place, so that differences of method would have minimal effect on correct distances.

The previously existing account of Scandia in which Johannis Magnus figured is unknown to us. Evidently it used the same kind of polar coordinate system used in the Vienna-Klosterneuburg corpus. Ziegler's successor and Johannis's younger brother, Olaus Magnus, published a similar but more accurately detailed map in 1539 (Figure 66). Along with it came an extensive textual description, Olaus's *Historia,* but he gave no further hints of its source.

Ziegler also propounded his own theory of why communication with Greenland had decayed;

> [Norway] was sometime a flourishing kingdom whose dominion comprehended Denmark, Friseland, and the islands far about, until the domestical empire was

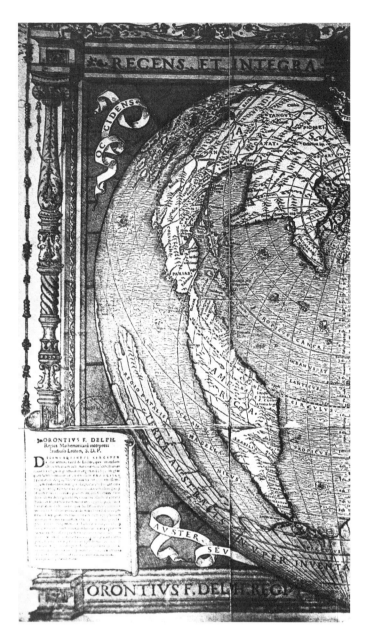

Figure 64. Cordiform planisphere by Orontius Finaeus, 1534–36. Location: Paris, Bibliothèque Nationale de France, Rés. Ge. DD.2987. Source: Lucien Gallois, *De Orontio Finaeo* (Paris, 1890), pl. 1. Courtesy of Map Division, The New York Public Library, Astor, Lenox and Tilden Foundations.

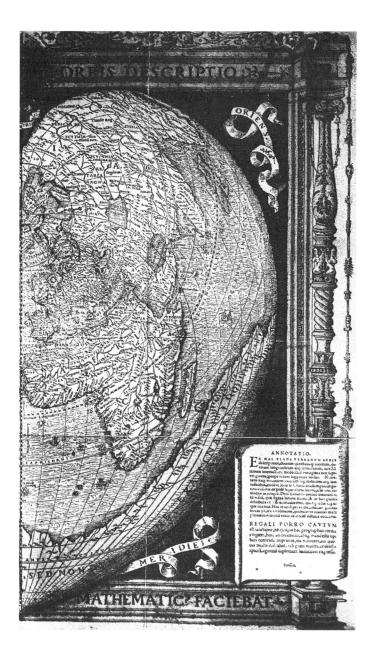

governed by the succession of inheritance. In the meantime, while this governance ceased for lack of due issue, it was instituted by consent of the Nobility that the kings should be admitted by election: supposing that they would with more equity execute that office forasmuch as they were placed in the same by such authority, and not by obtaining the kingdom by fortune and new advancement. But it came so to pass, that as every one of them excelled in richness, ambition and favor by consanguinity, so were they in greater hope to obtain the

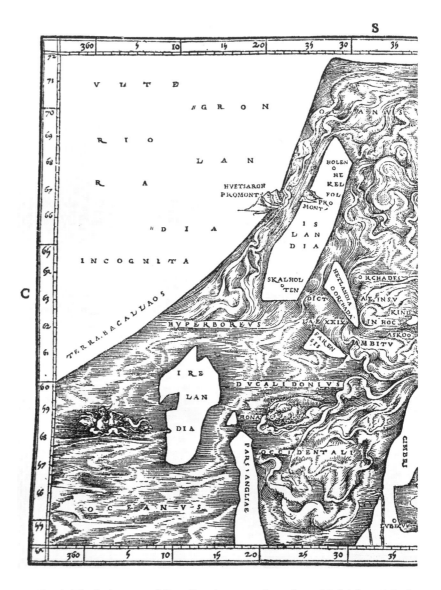

Figure 65. Jacob Ziegler's map of Scandinavia, 1532. Locations: Multiple originals. Source: Nordenskiöld, *Facsimile Atlas*, p. 57, fig. 31. Courtesy of Map Division, The New York Public Library, Astor, Lenox and Tilden Foundations.

kingdom; and were by this means divided into factions, attempting also occasions to invade foreign realms whereby they might strengthen their parties. [Norway] is therefore at this present under the dominion of the Danes; who do not only exact intolerable tributes, but also bring all their riches and commodities into Denmark, constituting the continuance of their governance in the infirmity and poverty of the subjects: which example, some other princes do follow at this day in the Christian Empire. . . . This is the fortune of Norway, whose edifices, towns

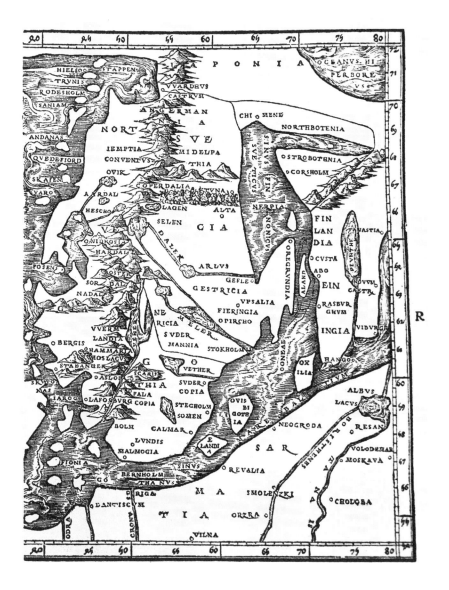

and cities can not defend their ancient amplitude and dignity: neither is there any hope of repairing their state.* For there are no consultations admitted for the redress of the commonwealth: No man dare show his advice or attempt any thing, uncertain of the minds and consent of others. To this difficulty is added the quality of the place. For the Danes have in their power all the navigations of Norway, whereby it may exercise no trade by sea, neither carry forth wares to other places.[4]

---

*Norway did not regain her status as a permanently separate nation again until the twentieth century. Explorer Fridtjof Nansen was a moving force in the reestablishment.

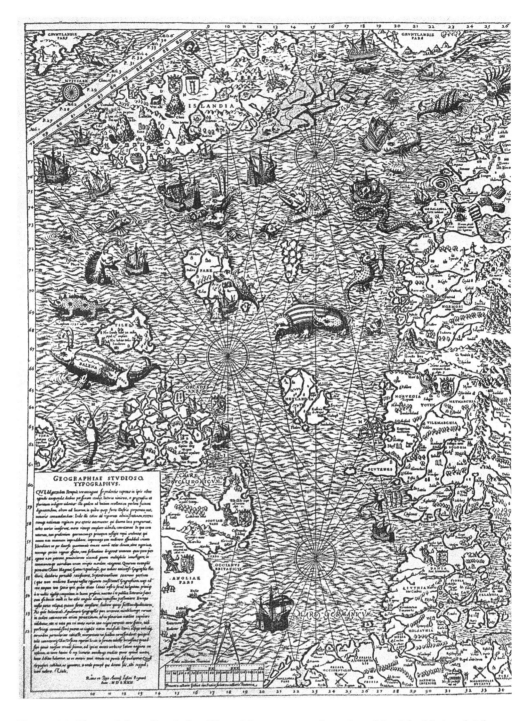

Figure 66. Chart of Scandinavia by Olaus Magnus, 1539. Location: Munich, National Library. Source: (1572 Rome copy) Nordenskiöld, *Facsimile Atlas,* p. 59, fig. 32. Courtesy of Map Division,

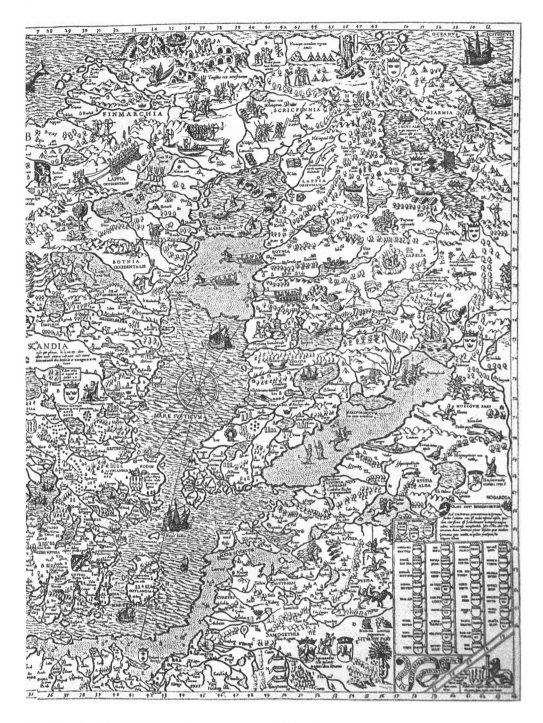

The New York Public Library, Astor, Lenox and Tilden Foundations.

The apparent occurrence of the Canadian Arctic coast in detail on several maps of this period corroborates its appearance on the Behaim globe.

A completely different area to summarize involves theoretical Norse cartography. The Scandinavian Norsemen before Magnus and Aorosien had no history whatsoever of mapping their own shores. We have seen that the American Norsemen met with and perhaps learned how to make maps from the Eskimos in earlier centuries. This seems to have occurred just at the time when the Greenlanders would have been learning how to use paper (or parchment) from the Icelanders. We must look upon any such new mapmaking capability of the Greenland Norsemen as an art form or technology unique to them alone among the Scandinavian descendants. It would have been a defining part of their culture. The first known map showing Greenland's outline in an informed depiction is the Yale Vinland Map (if it is authentic). On it also occurs an informed rendition of Iceland that could only have been made by a member of the culture that made the Greenland map possible. Native Icelandic mapmaking under European influence did not otherwise occur until the end of the sixteenth century. This could imply either that the Yale Map is inauthentic or that the mapmaking culture was diffusing eastward towards Europe. It is not inconceivable that it was the same culture that appeared in mainland Scandinavia when Johannis Magnus's informed data on Scandinavia came into Ziegler's hands. So informed, native Scandinavian cartography was apparently born full-blown.

CHAPTER 16

THE MISUNDERSTANDINGS ARE RESOLVED

AS THE MODERN PICTURE OF WORLD GEOGRAPHY BECAME CLARIFIED, scholars began to think in a new paradigm. They also came to realize that various ancient stories of travels to strange lands might be true accounts of bygone encounters with the New World. As a result, various geographical materials came to the surface that might otherwise have been lost to the historical record.

1558: Nicolo Zeno the Younger. It may fairly be said that the subject of the Zeno voyages (see pages 138 and 143) was a controversial one until Frederick Lucas published his overwhelming critical attack in 1898. After that it was a discredited subject. Lucas's attack on Zeno was even more virulent than the McCrones' attack on the Vinland Map at Yale. Lucas's method was to show that many portions of the Zeno narrative are highly similar to certain post-Columbian narratives on American lands published prior to Zeno's. He also criticized Zeno's map (Figure 67) for including segments based on earlier published maps.

   In spite of many subsequent writers' reluctance to concern themselves with Zeno following this indictment, one must admit that he still has not had a complete, fair trial. Indeed, most of the maps we have looked at included segments based on prior maps. Also, if different narratives on the same subject are true, then they *should* contain similar subject matter. At least one highly respected modern writer, E. G. R. Taylor, has concluded that this narrative makes sense as a true story. Recently, scholars are coming more and more to accept these voyages as authentic, even while acknowledging some plagiarism.[1]

   If even one item of credible information can be found in Zeno not at-

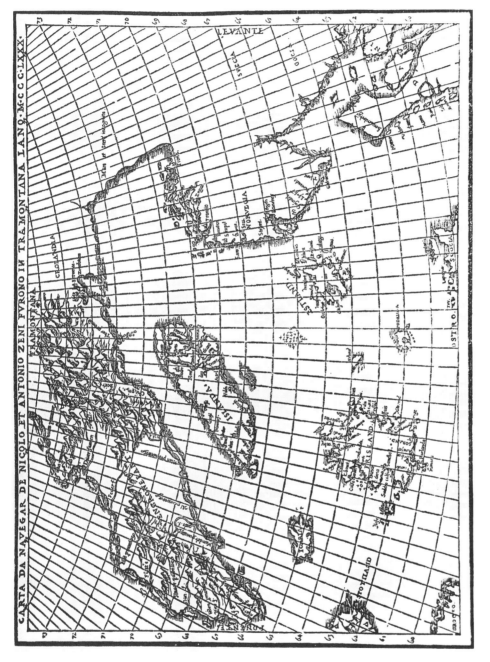

Figure 67. Map by Nicolo Zeno, 1558. Location: Multiple originals. Source: Nordenskiöld, *Facsimile Atlas*, p. 53, fig. 29. Courtesy of Map Division, The New York Public Library, Astor, Lenox and Tilden Foundations.

tributable to preceding writers, then his case deserves to be reopened, at least on that item. His detailed description of a kayak is unprecedented,* as is his description of Thule arch-wall construction. Another such item is the unique bailing sleeve pump on the inside of a kayak (see page 145). This is unfamiliar to experts on the history of kayaks, but it might represent a forgotten characteristic of the now defunct Thule culture. The two pieces of wood could be used in the manner of an old-fashioned washing machine wringer. This would drive the water out of the sleeve after its top opening had been made secure and a tie around its bottom opened. Such a device could have been useful in a kayak. The necessity of undoing the waist cowling to accomplish standard bailing would expose the interior to further swamping in heavy weather and defeat the entire purpose of the kayak. Instead, the occupant could simply withdraw his arms from his loose sleeves into his loose coat and safely go about using the bailing device by feel. (Of course, he would meanwhile be vulnerable to strong waves without balance control from his arms, but it still might be safer than undoing the cowling if no close-by ice floes offered haven.) If Zeno got these particular items from his ancestor rather than by plagiarism, as seems necessarily the case, then we can be sure that his ancestor must have been in Greenland as claimed.†

As for Zeno's map, he did not actually claim it had been made in its entirety by his ancestors. He merely suggested that he had extracted the essence of his ancestors' map into his own. Indeed, his whole purpose in supplying the readers of his narrative with a map was to "serve as a light to make intelligible that which, without it, they would not be so well able to understand." With this aim, his best procedure would have been to superimpose whatever he extracted from the ancestral map onto a modern map. That is just what he seems to have done. However, in spite of his claim, "I have succeeded in doing tolerably well," he has actually created more confusion than understanding.

In merging his two or more sources he has mistakenly depicted Iceland twice. His "Frisland" comes from another map, which, according to Lucas, represents Iceland's coastline. The island Zeno labeled "Islanda" is copied ei-

---

*While Clavus alluded to kayaks a century before, he described them only as "hide boats."

†Herewith a few other objections to Lucas's attack (references to page and line in Lucas): To equate the map with the text is a grave error, in light of the duplication of Icelands—see text below (72, lines 10–12). Zeno does not actually claim the existence of a volcano. An erupting mud pit of some sort is possible. These are common in the vicinity of thermal springs and may well have provided unusual building materials (73, lines 30–35; 77, lines 22–26). Magnetic variation makes the direction closer to compass north (77, lines 30–35).

ther directly or indirectly from Iceland on Olaus Magnus's map of 1539 (Figure 66). However, Magnus's floes of pack ice off the northeast coast have in Zeno become islands. Zeno's geographical naïveté seems almost incredible, unless one postulates intermediate copying between Magnus and Zeno. I believe that I have seen a map predating Zeno's that also showed Magnus's ice floes deteriorated into islands, but I have been unable to relocate it. A most intriguing aspect of these islands is the names Zeno gives them. They are the same as those given similar islands on the 1450 Catalan map in Modena (Figure 40), which I have surmised to represent the Arctic Archipelago. It now becomes clear that Zeno's map has at least one tie with mapmaking tradition, which was more than a century old in his time and which was associated with the North.

Thus critics may be too hasty in their claims that Zeno's Greenland was simply copied from maps such as the Cantino map of 1502 (Figure 51). True, the southern portions of the two are nearly identical, but that would necessarily be the case for any two maps that represent Greenland as accurately as these two do. Likewise, Zeno's representation of the west coast, allegedly copied from the A-type Clavus-inspired maps, combines with the lower part to give greater accuracy than either alleged source. It is rivaled only by the 1440 Yale Vinland Map (Figure 36). It is difficult to accuse Zeno of inept plagiarism at one time and superbly insightful plagiarism at another.

Plagiarism was not only common in Zeno's time but standard. The unacknowledged incorporation of other people's work into one's own was as ordinary as finding a bibliography in today's books. Zeno's cartographic entities of Icaria and Estotiland seem to be derived from the Olaus Magnus entities of Tile and Iona, while his Estland comes from the Shetlands. I cannot suggest any previous cartographic inspiration for "Drogeo" other than the New World identification (see page 238).

Zeno's narrative also undoubtedly borrowed some material from other writers to replace the substance of the family manuscript which he says he destroyed as a child. As early as 1536 his relative Marco Barbaro attested to the existence of a Zeno family tradition of the voyage to the North.[2] To the extent that that narrative contains unique material which cannot be exactly attributed to specific other writers, it still serves as a window into the past. Claims on the basis of Zeno for a Venetian discovery of America have surely been discredited, but the study of his publication should not be discouraged. It is time to put away Victorian attitudes of historical paranoia, vindictiveness, and zealotry and to approach Zeno with cold scholarship.[3]

1569: Gerard Mercator. In the Arctic portion of his world map of 1569 (Figure 68a and b) Flemish geographer Gerard Mercator copied his depiction of Greenland from Zeno. He has departed from previous tradition, however, by showing Greenland as a separate island rather than a mainland peninsula. This information presumably came from the *Inventio Fortunatae.* This seems consistent with Clavus's sources, for even though Clavus's pictorial map showed a peninsular Greenland, his tabular data referred to it as an island.[*]

Turning to Mercator's Septentrional Isles, it is clear that his representation shows influence from the Ruysch-Schöner-Finaeus school. However, it is also clear that his representation is not a copy of the previous series, which in particular did not contain "Grocland." (Ruysch did show the small split island in the same position as Grocland.) Although Mercator's "Groenlant" is a copy of Zeno's representation, if we look a little closer at the Septentrional Isle bearing the label "Pygmaei" we may be seeing still another representation of Greenland. The outer coast of this island is quite different from the other three and seems a fair representation of the west coast of Greenland. There is even a suggestion of the narrow glacier-free strip along Greenland's agricultural west coast, in that the encircling mountain range recedes from the coast there and only there.[4]

1570: Abraham Ortelius. Ortelius was a very close friend and countryman of Mercator's. His publication of an atlas in 1570 was the culmination of many years' work searching for all available sources. His map of Asian Tartaria (Figure 69) apparently shows the same Ungava area of the Quebec peninsula as does the Schöner-Finaeus series (see page 265), and in the northernmost peninsula an inscription says: "Danish sector, undergoing decline and evacuation." In those times *Danish* meant "Scandinavian." So this is the first surviving document that without equivocation places the Norsemen on the continental map of eastern Asia.

This is direct evidence for the actuality of the "Grand Misunderstanding" under the one-ocean paradigm in earlier times. The otherwise rampant misunderstanding about a drift away from Christianity is refuted by another inscription: "Region of the Nepthalites. Nepthalites from the territory of 10

---

[*]Aside from the Yale Vinland Map of 1440, the only other post-Clavus maps that make Greenland an island are one drawn in 1552 by John Dee (an acquaintance of Mercator's), Mercator's own map of Europe drawn in 1554, and a map by Humphrey Gilbert (an acquaintance of Dee's).

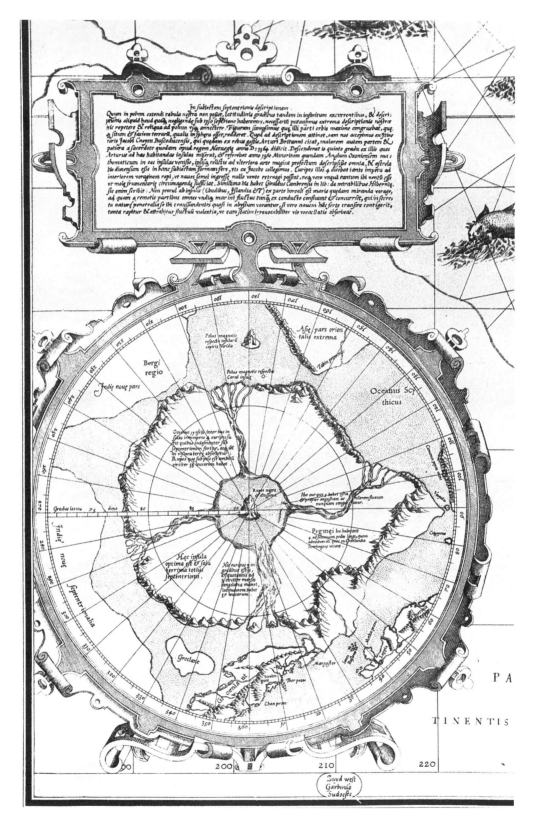

Figure 68(a). North Polar insert of Gerard Mercator's world map, 1569. Locations: Paris, Basle, Rotterdam. Source: Jomard, *Monuments,* no. 79. Courtesy of Map Division, The New York Public Library, Astor, Lenox and Tilden Foundations.

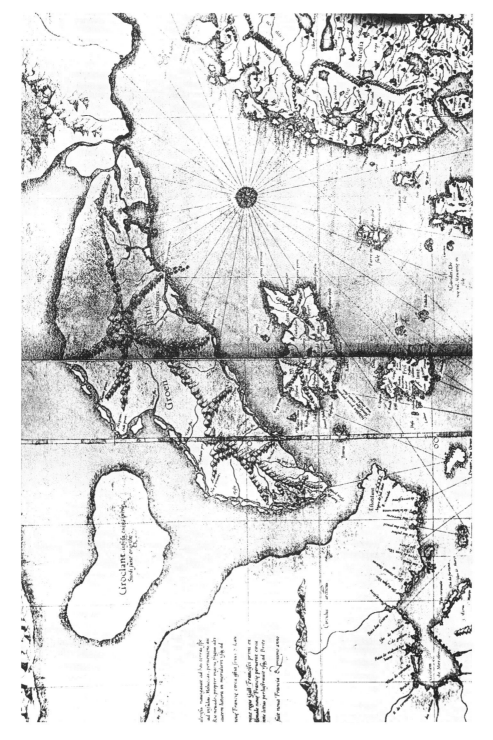

Figure 68(b). Detail showing Mercator's information on the Greenland-Grocland area. Source: Frederick W. Lucas, *Annals of the Voyages of the Brothers Zeno* (London, 1898), pl. 13.

Figure 69. Abraham Ortelius's map of Tartary, 1570. Location: Multiple originals. Source: *Theatrum Orbis Terrarum*, 1964.

tribes known by the abbreviated name 'Neptali,' behind the Danes, whom others incorrectly call 'Euthalites' and who in correctness should be said to be the Northern Danes, on account of the vaguely redeeming rule that they are bowed in subordination to Christianity and in the 476th sumptuous year

of having been conquerors of evil." One can overlook Ortelius's preoccupation with the lost tribes of Israel. The 476 years of Christianity before 1570 comes out to A.D. 1094. This year lies between the Christianization of Greenland (ca. 1000) and the voyage of Bishop Eric Gnupsson in search of Vinland (1117–21). It may reflect some record of a Christian settlement on the mainland. Ortelius's belief that the "Danes" had been in eastern Asia existed simultaneously with a growing suspicion that North America could *not* be Asia, because of the size of a degree. This would imply that the misunderstanding of the Danes in Asia was not Ortelius's own but predated him.

Regarding the continental separation, an erroneous report by Verrazano in 1524 had given rise to belief in an extension of the Pacific occupying all of what is now the interior of North America. This led to a departure from the Schönerian school, showing the New World as continentally separated from Asia by a vast "Sea of Verrazano." However, Ortelius, along with Mercator and a few other predecessors, narrowed this separation to a narrow strait. This "Stretto di Anian" is reminiscent of the 1530 sketch map (Figure 59) and Plethon's "great arctic strait." The actual existence of such a strait was not demonstrated to the modern world until the eighteenth century, by the expeditions of Vitus Bering. Perhaps whatever source informed Ortelius about the Danes reaching "Asia" (before the reality of the intervening New World was known) also contained traces of Eskimo information about the Bering Strait.[5]

Circa 1600: Icelandic Vinland Maps. Awareness of the geographical situation of northern America was increasing, and a more settled political configuration was developing in Scandinavia. The latter produced the beginning of the modern era of interest in the old Norse settlements west of Iceland. Denmark had come to be the dominant power in Scandinavia and was now assuming the duty of reestablishing awareness of the old settlements. There was both actual exploration and research into old documents. There gradually accumulated in the museums of Denmark invaluable collections of original documents from Iceland. Their contents must have held the keys to many of the mysteries that still confound us. I say "must have" because, by one of the calamities of history, most of these documents were destroyed when the Copenhagen museum burned in the next century. Many others, retained at the Skálholt cathedral in Iceland, were also lost to fire.

Two documents escaping such a fate are a map by a young Icelandic scholar, Sigurdur Stefánsson, drawn about 1590 (Figure 70) and a map by the Icelandic bishop Hans Poulson Resen drawn in 1605 (Figure 71). Comparison of their depictions of Greenland and the features south and west of Greenland reveals that they worked from related sources, if not the same source. However, there is a 90° difference in the two cartographers' orientations of north. Knowledge of Stefánsson's map exists only through copies

Figure 70. Anonymously drawn 1669 copy of map of Greenland and Vinland made by Sigurdur Stefánsson in 1590. Location: Copenhagen, Royal Library, G.K.S. 2881. Source: Skelton et al., *Vinland Map*, pl. 17. Courtesy of the Royal Library, Copenhagen, and Yale University Press.

made in 1669 before the fire, and the copyist (who incorrectly copied the date as 1570) had no knowledge of Stefánsson's sources.[*] Resen's original has escaped loss, and in its title he wrote, "The appearance of Greenland and

---

[*]There is some controversy about this date. There is known to have been another Sigurdur Stefánsson, alive at Skálholt in 1570.

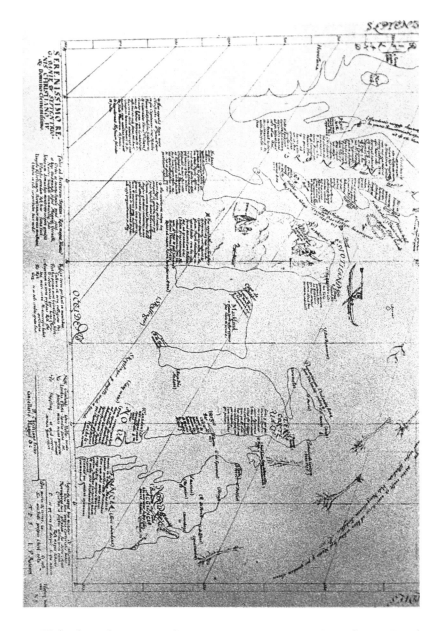

Figure 71. Vinland map by Hans Poulson Resen, 1605. Location: Copenhagen, Royal Library, Map Room. Source: Skelton et al., *Vinland Map,* pl. 19. Courtesy of Royal Library, Copenhagen, and Yale University Press.

neighboring regions to the north and west, according to an ancient sketch-map some hundred years old, from Iceland."

An ongoing controversy surrounds the phrase "some hundred years old." The original Latin is "*ante aliquot centenos annos,*" which has occasionally been translated "some centuries old." But Peter Hogg showed that the word *cen-*

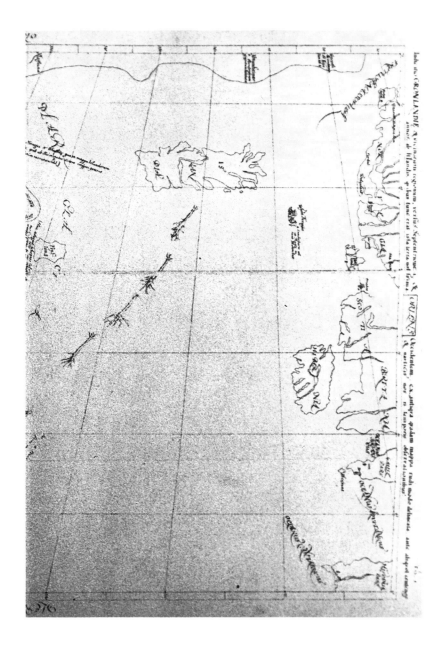

tenos is properly a cardinal number modifying *annos* and does not translate as "century." However, instead of interpreting *aliquot* as "some," as do all dictionaries, Hogg interprets it as "about" in the sense of "nearly."[6] His entire reason for making this assumption is to place the unknown prototype before Columbus, so that he can draw some vague parallels with known post-Columbian maps. However, the concept "some hundred years" is almost always interpreted to mean a hundred plus some, while a hundred minus some is usually expressed as "about," "nearly," or "around." Since we have seen much evidence of pre-Columbian cartographic activity in the Norse world, there is no reason to be forced into a somewhat unnatural interpretation.

Let us guess at an identification of the features in Resen's and Stefánsson's maps.[7] Clearly, neither of them is an accurate representation of any actual features west or southwest of Greenland. The two prominences labeled "Helluland" and "Markland" do not exist anywhere, regardless of latitude, in the actual North American mainland. But on a somewhat reduced scale, as seen in Stefánsson, they are suggestive of the peninsulas of Baffin Island. The sharp north-pointing "Promontorium Vinlandia" is well situated to represent the Cape Chidley Peninsula. Even a narrowed Ungava Bay appears to be present just to its left. I believed this to be exactly the true location of Leif Erikson's Vinland, as I argued in *Viking America*. The problem with this identification, of course, is that Baffin Island has three prominences rather than two. It turns out that the very fact that the prominence is missing may lead to further insight.

Let us think about the lost Icelandic prototype behind these maps, some hundred years old around 1600. This puts it in a fair position to be indirectly related to the 1440 Yale Vinland Map. In fact, that very map, presumed also to show Baffin Island, also contains an anomaly regarding this prominence: the inlets of Frobisher Bay and Cumberland Sound are fused into one, so that the three prominences appear as two. We see a similar fusion in Figures 8 and 9, in the component regional maps 2 and 3, based on local knowledge. Such regional maps could also have been the grandparents of the Icelandic vinland maps. During our analysis of the Yale map we speculated on the possibility that information from the *Inventio Fortunatae* could have gotten into the Yale Vinland Map. Hogg thought he perceived some resemblance between these Icelandic maps' Greenland and the Greenland on the 1507–8 Ruysch map, which we know was influenced by the *Inventio*. Aside from information that may have been transmitted via Ruysch, wisps of information from the *Inventio Fortunatae* may still have had an indirect input to these Icelandic maps.

SUMMARY

Apparently, by the middle of the sixteenth century, information about contemporary or past Norse contacts with some unknown distant land no longer required accommodation by geographers by means of mistaken paradigms. The scholars then turned to the retrospective study of Norse historical material and laid the foundations for subsequent inquiry into the process of the discovery of America.

CHAPTER 17

CONCLUSION

ONE UNDERLYING PURPOSE OF THE CHRONOLOGICAL SURVEY was to provide an exhaustive test and list of evidence for the hypothesis of Native American-to-Norse-to-European information transfer under the one-ocean paradigm and the provenance paradigm. The test was to see whether that hypothesis (with others) could account in a plausible way for various cartographical and documentary oddities in the geographical record of the North. Were the five criteria in our coastline identification rule in Chapter 1 adequate? Did we ever bend the rule too much? Was there indeed a cartographic iceberg body under the cartographic peaks we have seen? We have demonstrated a book full of expectable if not exactly necessary conditions here, conditions that could be expected if the hypotheses were true. This approach to the hypothetico-deductive method is useful when exact deductive predictions cannot be made. Does the evidence we saw constitute a smoking gun or just smoke and mirrors?

The enthusiast may wish to conclude that our entire view of history should be changed. The traditionalist may wish to conclude that the result is merely a fantasy illustrating the potential for accidental similarities of coastlines, or may even claim that there in fact were no similarities. Any individual piece of evidence examined here, viewed in isolation, would be flimsy indeed. Nevertheless, perhaps either extreme camp would be jumping to a conclusion without further evidence and analysis.

An analogy can be made here between our situation in historical cartography and a situation in historical linguistics. When the existence of an Indo-European group of languages deriving from a single lost parent language was first postulated in the nineteenth century, it was based on perceived similarities of cognates in the supposedly related languages, along with pronunciation-shift paradigms. Skeptics rejected these similarities as accidental, but further research soon made a convincing theory. The same method of cognate similarities eventually uncovered other language classes, apparently deriving from unique parents: Afro-Asiatic, Dravidian, Uralic, Altaic. Again, accident was claimed, but the theories prevailed. Most recently, a new theory using related similarity methods has postulated a single ancient language, dubbed Nostratic, underlying all of these families. Again accident is claimed, but positive evidence is accumulating. In all these cases, the tide of skepticism was finally turned by the uncovering of fixed paradigms that encompassed the observed similarities and changes.

The least that can be said of the results of our survey is that the idea of a hidden divulgence of America no longer sounds as implausible as it might have in the beginning. A second reading (after a suitable rest period), better to absorb cross-referenced corroborations, might make it seem even more plausible. Admittedly, there are numerous maps of the same period, not examined here, that show no traces whatsoever of the kind of evidence we looked at. But these may be construed as examples of the principle stated in the introduction that material that cartographers could not accommodate to was they simply ignored and discarded. Other world maps of the period contain variations of the surveyed material, but many of those are clearly derivative from information in the maps or sources we examined. I have searched among all of the better-known and many of the lesser-known collections of facsimile maps as well as many originals in several of the best libraries of the world. I have attempted to explain every map I could find that preceded the post-Columbian Arctic explorations and had some anomalous or fantastic feature in its northern parts. I am not aware of any documents that disprove the hypothetico-deductive structure proposed here.

Nevertheless, failure to find a counterexample does not constitute proof in any absolute sense. Indeed, some with a traditional philosophical inclination may wish to dismiss our approach as "advocacy science," wherein a hypothesis goes seeking evidence instead of the evidence forcing conclusions. But the former is how most progress is made in science nowadays, unaided serendipity being rare. It is possible to find alternate explanations of some of the geographical oddities at issue, or even different coastal interpretations, as has been done in some of the individual references cited. Those alternate explanations, however, are isolated cases with no common underlying paradigmatic structure; and when gathered into a survey they would seem to take on an ad hoc character, appearing to be a means of explaining away the oddities. Our explanation is the most parsimonious.

As for the argument that the coastline similarities we saw could be accidental under the law of large numbers, it must be admitted that the number of medieval maps originally created, let alone preserved until today, does not constitute a truly large number. It may also be argued that some of the similarities might be the result of the fractal self-similarity of coastlines, if they are not corroborated by other evidence. But in the real world the scale changes necessary to find such correspondences are many powers of ten, and even then the similarities are isolated. The few scale disparities entertained here have been much less than that. Concerning fractal scaling Mandelbrot says, "Here as in standard geometry of nature, no one believes that the world is strictly homogeneous or scaling."[1] The largest scale adjustment we used, in Clavus's first map, was one power of ten, and there are numerous reasons to doubt accident there; the others, in Clavus's second map and in Behaim's globe, are comparable with historically acknowledged scale disparities cited in Scandinavia on the Behaim, Dalorto-Dulcert, and de Virga maps, in Greenland on the Vinland Map, and in the New World on the La Cosa map. Only two consequential rotations were invoked. The one in the second Clavus map was thoroughly justified in the context of Eskimo orientation procedures. The one in Contarini's and La Cosa's was justified only by its results.

It would seem promising to attempt a statistical analysis of our situation. However, the theoretical statistics of such coincidences are not as well established as one might think. Charles Hapgood used statistics in a very aggressive way trying to prove his interpretation of part of the Piri Reis map as Antarctica. Many have nevertheless rejected his thesis as implausible,[2] and the main result of his work seems to this author a demonstration that statistics prove nothing (or anything) regarding old map interpretation. More recently statisticians at Harvard University have endeavored to lay a foundation for a theoretical study of accidental coincidences.[3] Their results confirm what we already sensed: coincidences occur in the minds of observers. We may try to sway those minds with familiar methods of argumentation, but only in exceptional cases can we hope to find objective reality through statistics. Another tool that may come to mind is the field of map accuracy analysis, such as we used in Chapter 2 for the first Clavus map. This is also essentially a statistical method. One kind of mathematical device that would seem useful is what engineers call the "signal to noise ratio," to give an idea of meaningful coastline correspondences versus accidental correspondences. In excessively "noisy" situations, engineers can sift out meaningful signals by examining the largest possible sets of data and looking for the most persistent signal. My analogous approach has been to examine the totality of still-existing medieval maps and confirm that the Arctic anomalies all fit into the paradigms, without exception. In the end, we are left with the human brain's pattern-recognizing ability. The same brain circuitry that recognizes map shapes as familiar and identifiable is the one that has evolved to recognize and identify human faces, their relatives, and ethnic classes. It is difficult to specify how it does this, but it is effective. We can expect it to create some consensus regarding these maps.

While the evidence here is imperfect, it is not insignificant. It is particularly difficult to believe that the similarities to American lands in the Clavus maps, for example, could be accidental. Their interpretations are supported by topologically separate interior features and textual corroboration. It is also difficult to find convincing alternate explanations of several other suggestions of the correctness of our hypotheses which include textual corroboration: de Virga's map, the Yale Vinland Map, La Sale's description, the Vienna Klosterneuburg Schyfkarte, Schöner. De Virga and Ortelius both gave independent direct, deductive evidence of the Grand Misunderstanding from both east and west.

Many of the suggested graphic correspondences corroborate one another. Probably the most uniquely characterizable piece of mainland west of Greenland is the Quebec-Labrador peninsula. It was identifiable in eight different maps, each with uniquely different information.[*] Another unique feature of North America is Hudson Bay, which also helps define the Quebec/Labrador peninsula. It or unique information about it can be seen in four relatively independent sources.[†] Of course the most unique feature of Arctic North America is the Arctic Archipelago, which was discernable in seven different depictions.[‡] Other identifications are corroborated in twos, threes, and fours in the maps we have looked at. Besides these maps there exist scores of chronologically later maps, by cartographers not examined here, that display similar features. Some of these may involve further influence of the American prototypes as well as exhibiting lineal descendance.

It is understandable that, given the involvement of bygone perceptions, paradigms, and misunderstandings in the thesis, hard evidence might be elusive. In fact, contemporary research into the nature of scientific progress leads

---

[*]That of de Virga, Genoese map in Florence, those of Fra Mauro, Ulm Ptolemy, La Cosa, Contarini, Schöner 1515, and the Icelandic Vinland maps, plus all of their derivatives.

[†]Fra Mauro, Behaim, Ruysch, Schöner (1515).

[‡]Dalorto-Dulcert, *Inventio Fortunatae,* Catalan Fragment, Vienna Klosterneuburg *Septem Climatum,* Catalan in Florence, Catalan in Modena, Behaim.

some to think that even in the natural sciences the absolute truth of a theory can inherently never be established.[4] A scientific theory can become increasingly probable, but that may be the best that can be hoped for. With induction it is always theoretically conceivable that some totally unexpected phenomenon will explain all the supporting evidence otherwise. Inductive methods can never establish absolute truth,[5] but people long for absolute truth in historical matters that are still being actively taught in schools. While it is not possible herein to provide absolute truth, there may be ways to deny absolute truth, to refute the theory. Perhaps one way to deny the induction given here would be to provide convincing, non–ad hoc alternate explanations for all the items in our survey. Another way, for those who still hold to accidental shape correspondences, would be to produce a contrary but similarly effective body of correspondences located somewhere else in the world under some equally likely paradigms.

Progress is possible even in the absence of absolute confirmation; witness the natural sciences. New theories become at least tentatively established on the purely utilitarian basis of their helpfulness in simplifying old mysteries in the field, on their prediction of unsuspected new phenomena, their aesthetic elegance, and even their promise of results for future research. In the field of jurisprudence, circumstantial evidence is completely valid and acceptable if it is sufficiently corroborated. One must decide if any remaining doubt in one's mind is reasonable. Darwin's evolutionary theory is neither deductively nor inductively proved, but is generally accepted on what are essentially overwhelming circumstantial grounds: its explanatory power. It is, in the end, not so much a scientific theory as a historical theory.[6]

The same attitude can be held toward the total theory set forth in this volume. Indeed, opponents can construe individual examples, both here and in evolution, to be far-fetched in isolation, but in their totality the theories have a great strength. Here they involve a wide variety of independent evidences, including not only maps but archaeological, anthropological, and linguistic considerations as well as textual documents. And no contradictory evidence has been uncovered. Such corroboration often suggests that a theory is, at least in its major respects, substantially correct, even if one cannot attach a definite probability value. Nevertheless, the total amount of available evidence, pro or con, that one has to work with regarding medieval cartography falls far short of evolution's test bed: the entirety of life. One assurance of a theory's correctness is whether it can make predictions for the future that are borne out by experiment. While such predictions for scientific theories are usually of a mathematical nature, one form of prediction that is testable, even if not experimentable, can be made here too. That is that previously unnoticed texts or maps will be found, materials that have been mystifying in the past and ignored as nonsense but which make elegant sense in the framework of this theory. At minimum, this would increase the assurance of correctness.

Another possible evaluative approach is to compare the proposed theory's success with previous explanations of the data observed. Namely, it would seem disingenuous at this point to go back to the old general view of coastal shapes: "If we don't understand it, it must be just a quaint fantasy." The present theory is much more efficient at explaining these oddities. The explanations given herein apply with a generality that, according to the principle of parsimony and the principle of convergence of evidence, stirs confidence. That is the best criterion of scientific validity.[7]

So, it can be said with a high degree of confidence that the theory has already been strongly corroborated and is probably true. It is more than a suspicion, and if it still remains a belief, it is a well-founded belief. Its explanatory power is indeed overwhelming. Yes, there probably were New World information sources tending to divulge American land, sources which were countered by the hiding effect of the one-ocean and provenance paradigms.* Serendipity could still uncover some deductive proof, now that attention has been focused. Perhaps this focusing of attention will be the major value of the theory. Some future independent discoveries that provide corroboration or refutation will let us know, perhaps fairly quickly.

All this being said, I will immediately acknowledge that the work presented here is only a starting place and leaves much to be done by others. There may well have been as many questions raised as answered. These are the fundamental theses from which further elaboration or testing of the theory might proceed:

- —There in fact was a late medieval and Early Renaissance Greenlandic reencounter and exploration of America, separate from the Eriksonian discovery of Vinland.
- —The Greenland Norsemen contacted the native peoples of America on a sometimes amicable basis. They engaged in cultural interchange in both directions with them and moved physically within their lands.
- —Most medieval and Early Renaissance European geographical materi-

---

*Even so, those sources remain exceptional. There is no suggestion of widespread Eskimo cultural transfer to the Norse.

als regarding the Arctic and many regarding the Orient that have heretofore been considered fantasies are actually records of Norse contacts with America and/or its peoples.

—Norse (or Norse-transmitted) geographical data about America had a wide variety of profound effects upon European theoretical cosmology and geography.

Certainly all of these theses seem likely to provide opportunities for further elaboration and exploration. The last three points lead to an ironic twist concerning the Eskimos. Rundstrom has observed that in providing nineteenth-century explorers with maps later used to overrun their lands, the Inuit were unwitting conspirators in their own disenfranchisement.[8] Many cultures encountering outsiders tend to be secretive about their internal geography, apparently for good reasons. Meanwhile, McGhee has speculated that the very deterioration of the Thule culture that yielded the modern Inuit culture was caused by stresses from the post-Columbian European contact.[9] Perhaps, therefore, the Thules' sharing of American geographical information with the Norse (and thus Europeans) contained the seeds of their own cultural destruction.

The penultimate item in the list above is still open to the question of which maps could have derived from native American information and which from Norse exploration. The evidence so far leans toward a native American derivation of most of the information. Nevertheless, recall the distinctly European flavor of the descriptions in the *Inventio Fortunatae.* And recall that some of the information would require the Norsemen to have gone at least as far south as Indian territory. Knut Bergsland's investigations of Eskimo language found the Norse word for "sheep" or "goat" all the way from Labrador westward along Ungava Bay to the east coast of Hudson Bay. While he concedes that this could be explained by more recent Norwegian missionary contact in the area, this seems equivocal.[10]

A more reliable guide to how far the Norsemen went might be sought in the archaeological record. Controversial runestones found at Kensington, Minnesota, and at Spirit Pond, Maine, would have made a nice contribution in this regard.[11] But the dynamics of persuasion might go the other way: the existence of these maps lends credence to the stones instead of vice versa. Less controversial finds along the eastern seaboard have been a Norse penny in Brooklin, Maine, and a Thor's-hammer amulet in Niantic, Connecticut. But these could have reached their discovery sites by Native American trade routes, and indeed are associated archaeologically with native trade items. In the North, the early decades of the twentieth century saw the discovery of numerous archaeological traces in the central Arctic that were identified as Norse. In the closing decades these were widely reclassified as belonging to the Dorset Eskimos, the antecedents of the Thule. In the new century, Patricia Sutherland is finding compelling archaeological evidence that the Norse went at least as far west as Baffin Island in the thirteenth or fourteenth century.[12] The archaeology of the Northlands is still in its infancy, and it remains to be seen what might appear.

The last thesis above, since it involves changes in firmly established ideas (even prejudices), is the most likely to generate controversy and demands for further proof. One avenue of further textual support might be a closer look at all late medieval or Early Renaissance texts mentioning Hyperboreans (and other northern people, both legendary and actual). Another promising locus for further attention would be the transitions between the provenance paradigm and the one-ocean paradigm in dealing with extensive land areas. We will do just this here and now, making a detailed diversion that leads to surprising results concerning the question of continuity between the Norse and Columbian discoveries.

Reviewing the early centuries of Norse contact with America, it appears that information about the New World was assimilated under the one-ocean paradigm, primarily by northern sources and Adam of Bremen. In the thirteenth century there was contact with fresh Thule information about America (and perhaps some new but related Norse exploration). These new descriptions arose within a tradition of pragmatically reliable quantitative data of a different character from qualitative descriptions gathered previously. By 1300, with the Carignano chart, the provenance paradigm seems to have become the primary way to assimilate data about American lands into the European map. Perhaps this switch was somehow associated with the rise of the mariner's chart and its quantitative requirements. Those requirements could have exposed conflicts between the distance to the new lands and the theoretical distance across the ocean. When southern European cartographers adopted the provenance paradigm, they seem to have let the one-ocean paradigm become dormant. For a century and a half, through the middle of the fifteenth century, the map of Scandinavia assimilated American lands as large as the subcontinental Quebec-Labrador peninsula. After about 1450, these lands were incorporated into representations of eastern Asia, under a resurgence of the one-ocean paradigm.

What made it conceptually possible for southern and central European cartographers to take lands that had been shown in the Scandinavian sphere and place them in Asia? The only answer I can see is that they thought the Scandinavian sphere in the West included Asia. That is, they must have had the concept that Scandinavians had been to Asia or had contact with Asians. Ortelius showed explicit traces of this belief. Why does the bridge between the Eskimo migration and the European cartographers remain unseen? Today's professional cartographers seldom have their technical deliberations enter into the popular historical record, and we should expect the situation to have been even less so during the Middle Ages. Nevertheless, there may well be some hint of deliberation about this in the documentary record. One might expect such deliberation to be tied up with considerations about the distance to Asia, eastward or westward, as d'Ailly made explicit. Ever since the controversy about Columbus and the court scholars, the idea of Asia's being in the West has stimulated discussion. We tend to think that anyone who had such ideas before Columbus would stand out in the historical record, because we have made Columbus's idea seem so radical. But this expectation may be a self- delusion brought on by scholarship in worship of the Columbus creed. It may well be that some pre-Columbian savants thought of Asia as being in the West without making a big deal of it. It is a natural thing to think about, and it need raise no eyebrows—unless one is trying to acquire understandably conservative royal financing.* Indeed, the temporary reign of the provenance paradigm may have been a reluctant interim concession by cartographers to arguments that Asia could not be so close in the West.

So, by Albert von Szent-Gyorgy's definition, that "discovery consists of seeing what everybody has seen and thinking what nobody has thought," it may be said that there was a legitimate discovery of America in the mid-1400s, when the return of the one-ocean paradigm allowed geographers to think new thoughts about old information. People had been seeing these coastlines on paper for over two centuries, but only after 1450 did some dare to think they might represent Asia after all. That discovery at present is attributed not to any particular individual but to an entire calling of cartographers mulling over some anomalous data sources. They had encountered something they justifiably accepted as real, which was eventually called America after yet another Szent-Gyorgy kind of insight in the teens of the new century.

---

*It could have been counterproductive for Columbus to argue from any Norse evidence, even if, improbably, he knew any consciously. Any rationale for prior European claims to the land would surely have undercut his royal financing, making his voyage impossible.

With specific regard to Christopher Columbus, admittedly nothing herein creates an inescapable compulsion to view his enterprise in the structuralist terms discussed in the Introduction. So far the most nearly direct link between the Norse explorations and Columbus is the possible appearance of Markland in a map that some historians suppose to have been Toscanelli-inspired (Figure 41), Toscanelli being Columbus's adviser. One additional possible link has been given by Ardell Abrahamson, who claims a likely provenance of the Yale Vinland Map in Columbus family archives. He suggested that this very map may be the one carried by Columbus and his associate, Martin Alonso Pinzon.[13] Nevertheless, one might still try to maintain, as his sometimes doting son Ferdinand did, that Columbus's inspiration was independent of any predecessors. Ferdinand believed simply that, "one thing leading to another and starting a train of thought, the Admiral while in Portugal began to speculate that if the Portuguese could sail so far south [along Africa], it should be possible to sail as far westward, and that it was logical to expect to find land in that direction." But the vague "one thing leading to another" is just what Piagetian structuralism tries to understand in a more rigorous way. It is difficult to hide behind that vagueness any longer in the face of the almost inescapable evidence of earlier American information and evidence that it influenced many geographers of Columbus's era. If one accepts as real *any* of the cartographic traces we have described, then one must acknowledge that they did have an influence on European thought, conscious or not. Once the door is opened, it is difficult to shut it selectively. In any case, the stage is set for giving more meaning to the question raised in Chapter 1 about the *timeliness* of Columbus's idea of Asia in the West.

Along with his brother Bartholomew, Christopher Columbus spent many a day as a full-time professional cartographer during the period 1476–83, when his plan was hatching, as well as thereafter. Surely he was exposed to the same paradigmatic influences we have seen displayed by other cartographers. This was just after new examples of the provenance paradigm ceased in favor of placement of new lands in Asia under the one-ocean paradigm. Paradigms tend to sweep up everyone in their field. Indeed, there exists an unsigned circular mappemonde containing internal evidence of having been made under the direction of Christopher Columbus ca. 1492 (Figure 72). There is even some controversial evidence that this very map was prepared as part of Columbus's presentation to Ferdinand and Isabella.* Here, in a complete rejection of the Clavus tradition that had been followed by car-

---

*This mappemonde is actually a minor decoration on a much larger sailing chart.

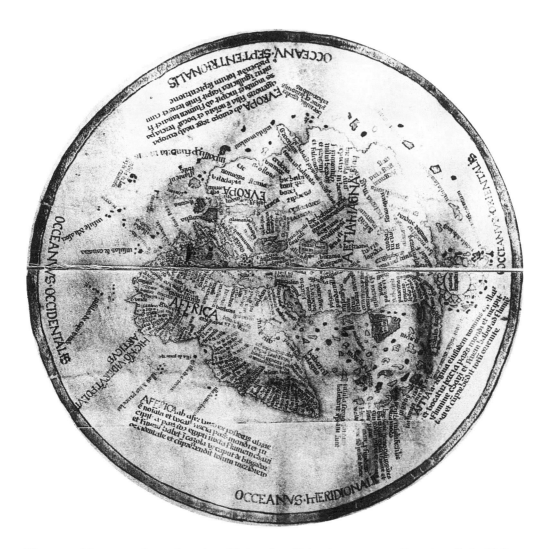

Figure 72. Mappamonde attributed to Christopher Columbus, ca. 1492. Location: Paris, Biblio-thèque Nationale de France, Geography Division, Ge. AA. 562. Source: Charles de la Roncier, *La carte de Christophe Columb* (Paris, 1924), pocket. Courtesy of Map Division, The New York Public Library, Astor, Lenox and Tilden Foundations.

tographers for sixty years, Greenland is shown not as a land arcing toward the west from east of Scandinavia but as a peninsula going straight outward from Arctic Asia.[14] On maps predating 1492, such a naturalistic and correctly pro-portioned outline of Greenland is surpassed only by the Yale Vinland Map. This depiction of Greenland as a projection from Orbis Terrarum is a revival of the old Norse idea of Greenland expressed in the days of Niklaus Bergs-son and the *Rymbegla* tract under the one-ocean paradigm. Under this idea, lands we today consider to be west of Greenland lie east of Greenland on the Asian coast.

Herein may lie the greatest significance of the long-standing Norse question of whether there was land under the Pole Star. If the Arctic shore of Orbis Terrarum falls short of the North Pole, then we are left with this contradiction between the theoretical mappemonde model and experienced reality of direction when sailing in the Greenland area and beyond. On the other hand, if land docs lie under the Pole Star, that is, if the pole is interior to the arc of Orbis Terrarum, then the theoretical model agrees with experienced reality. This is the model that would arise if the eastern seaboard of America were perceived as Asia, with the North Pole lying north of Greenland (and the polar region itself unknown). The theoretically eastern Asian lands would be reached by sailing westward from Greenland not very far from shore, regardless of the extent of the ocean, exactly as we envisioned Bergsson explaining the location of Vinland "The Good." This very situation may well have been the origin of the speculation that land must lie under the Pole Star (rather than any pointless musing about a yet unknown Arctic Ocean).

This opens a new dimension of possible meaning in Columbus's and Toscanelli's statement that by sailing westward one could reach the lands that are normally considered to be eastern. It could mean that the size of the ocean was only secondary to the fact that people had already gone there hugging coasts. The Christopher Columbus mappamonde leaves ambiguous whether the pole is interior or exterior to Orbis Terrarum. However, the tip of Greenland was known to point south, so the implication is that the pole was interior.[15] To anyone used to sailing in the Greenland area, it was clearly interior.

With this picture in mind, recall the quotation from the *King's Mirror* on the subject of Greenland: "lies on the extreme side of the world to the north, and he does not think there is land outside the disk of the world's land beyond Greenland [i.e., further outward], only the great ocean which runs round the world; and it is said by men who are wise that the strait through which the empty ocean flows comes in by Greenland." I have interpreted this to refer to Davis Strait and Hudson Strait. Just beyond Greenland on the Columbus mappemonde lies an embayment that would correspond to Davis Strait. It is not inconceivable that this represents Ginnungagap.

The Piagetian view of human nature at present offers the most hope that New World geographical history will progress beyond questions of idiosyncratic Columbian interest. One aspect of that view suggests that we should actually not expect to find traces of deliberation about the paradigm shift in the historical record. Such mutations are coming to be seen as spontaneous occurrences in self-organizing systems (such as the body of human knowledge) when they are forced to seek a new equilibrium.[16] If one accepts that view, then Columbus's activity may be regarded as a completely well-founded application (one of the first in his time) of the scientific method. First, he (hypothetically) became involved with certain observational data that was incompatible with existing theories (reports, direct or indirect, from recent Norse contacts). Second, he used scholarship and imagination to concoct a new theory able to embrace the deviant observations (his particular theory of Asia in the West). Third, he resorted to experimentation to prove his new theory (the ocean crossing). He never did get to participate in the fourth, recursive step—modification of interim theory on the basis of still newer observations. That step has also been denied many other people whom history nevertheless regards as pioneer scientists.

I do not suggest that Columbus necessarily went about these steps methodically, or was even aware that he was making them. His conscious mind was filled with religious and Ptolemaic rationalizations. These protoscientific steps would have been, as stated in the Introduction, "in the air." But I do suspect they were there, and their presence and influence deserve strong probing by historians of science. The astronomer and historian of science Owen Gingrich suggests that Columbus's voyage, in breaking the medieval geographical mind set, was ultimately responsible for the new thinking that led to Copernicus's ideas.[17] Similarly, the return to the one-ocean paradigm from the provenance paradigm might have helped loosen the medieval mind set and open the way for development of the scientific method. Further study of the transition from an Early Renaissance Norse encounter with American natives to the European exploration might extend and strengthen the rational as well as the psychological foundations of the history of science.

Modern readers steeped in our two-hemisphere paradigm* may find it difficult to see why the divulgence-hiding paradigms survived for so long. It is also difficult to see why it took until now for anyone to notice their traces in the cartographic record. But that is the very essence of such a paradigm in any field. Adherents go to any lengths to maintain the paradigm. The Norse-Eskimo divulgence of America has seemed hidden to us only because we had limited our ways of seeing. In this respect Columbus can be looked upon as an ultraconservative, one of the last of the old guard still framing his thinking in the Middle Ages. The new thinking that came along two decades after Columbus—the realization that there was a whole new land hemisphere—was so powerful that it makes it difficult even now to imagine the

---

*Some writers would refer to this as the terraqueous paradigm.

pre–New World mind. We are accustomed to thinking of Columbus's idea of sailing westward toward Asia patronizingly and with some arrogance, needing to excuse his theory. But our surmise here gives him good grounds for staking his own and other lives on the theory. Researchers able to put themselves into the pre-Columbian frame of mind will likely find answers to other mysteries in the historical record. Perhaps they will even find the "smoking gun" of evidence that turns my suggestions regarding Columbus and his predecessors into inescapable conclusions. This author has certainly left many stones unturned. No doubt the evidence is there, pro or con, if we just learn how to select the stone with the gem under it by asking the right questions.

If the surmise regarding Columbus is upheld, it may weigh in the perennial discussion of whether history proceeds from individual human actions or is a more complex phenomenon of flowing events propelling individual humans with it. Historians have put too much emphasis on the European discover of America as a single event accomplished by a single ethnic hero, whichever one. Instead, that discovery (or divulgence) should be seen as a protracted joint uncovering, a continuum starting with the Norsemen and proceeding over many centuries and involving several ethnic groups, all deserving credit. Today's vast development of the New World has roots in two interacting medieval migrations, one westward across the Atlantic and another eastward across the Arctic. Of course, there were many other factors involved, and it would have happened eventually anyway, but perhaps not quite when it did or with such fascination. After surveying this history, one may look at the map in the front of this book and sense that America was virtually trying, over the centuries, to get itself discovered.

In any case, America was clearly an intellectual as well as a geographical discovery. Its discovery went through many phases, from the paradigm that placed it in Scandinavia through the several paradigm shifts placing it in Asia through the voyages that explored it to the eventual realization that it was an unrecognized continent. Further understanding of this extended process of discovery will undoubtedly continue to emerge.

THE VINLAND MAP'S INK

THE INK OF THE YALE VINLAND MAP has become perhaps even more famous than the map's coastal outlines. In 1974 the Walter C. McCrone Laboratories of Chicago announced that it had found a modern form of anatase titanium dioxide inside the ink vehicle.[1] This white pigment was asserted to have a particle size distribution exactly matching that required for maximum visual light reflection, as manufactured in the twentieth century. Walter McCrone immediately came to the conclusion that there was no way other than twentieth century forgery that the anatase could have gotten into the ink. In this appendix I will demonstrate that there is at least one plausible way other than forgery.

McCrone's forgery scenario noted several other anomalous features. He described the pale, yellow-brown binder of the ink as "different from that of any other known medieval ink."[2] This fact by itself is not necessarily indicative of forgery. Medieval scribes mixed their own ink from privately or even secretly distributed recipes that had evolved by trial and error. A widely used ink was iron-gall ink, essentially tannic acid and iron sulphate. This had the strange property of being quite pale upon application and not turning fully black until after several days of oxidation. However, some scribes experimented with novel inks and even distributed recipes that were incorrect. One comparatively frequent error was to omit the iron-bearing component in iron-gall ink.[3] This omission could result from haste or negligence as well as from an incorrect recipe.

In fact, the Yale Vinland Map's ink does exhibit an extremely low iron content.[4] I have conducted a simple experiment in imitation of the omission of iron from an ink. I applied pure tannin to paper and let it age for several months. The result seems strikingly similar visually to the pale yellow-brown binder of the Vinland Map. Af-

---

This appendix includes material presented by the author at the Seventh International Congress on the History of Cartography, Washington, D.C., 1977. This material was the subject of a television interview with the author on the *Archaeology* series and was also posted to the Internet's "Maphist" e-mail list on January 9, 1997. It appears here for the first time in print.

ter the initial yellowing, its color remained static. This suggests that that binder may simply be gall extracts without iron salts. In the more than twenty years since I first made this suggestion, none of the principals in this controversy has suggested a test of the Vinland Map's ink to prove or disprove such a conjecture.

McCrone's forgery hypothesis is based on a second anomaly, in addition to the presence of the titanium dioxide pigments in the ink, namely, that there are microscopic flakes of a black carbon ink layer that once overlaid this presently exposed pale yellow-brown ink. He interprets this as a deliberate, painstaking attempt to simulate aged ink.[5] The shortcoming of this hypothesis is that it requires that a modern forger go to rather extreme lengths and yet amateurishly end up with a demonstrably anomalous layering. A different explanation could be that this black layer, which is now almost completely flaked away, it merely the remains of an imperfect restoration attempt sometime during the intervening centuries. It could even have resulted from a retracing by the original scribe after he realized that his original ink was not turning black. This case would account for the accuracy with which the restorer was able to overlay the original, because both would have been in the same hand. If his job depended on delivering a useable product, he would have had sufficient motivation to do the painstaking work.

The major anomaly of the ink is, of course, the titanium white pigment particles that permeate the ink's binder. Might there be an explanation other than forgery for how they got into the ink? McCrone apparently never considered such a possibility. He himself stated that the pigments do not actually contribute anything to the pale yellow-brown color of the ink; the color is determined completely by the binder and/or impurities, and the white pigments seem purposeless. It is true that experimental titanium pigments of the 1920s did have impure colors that match this ink binder's color, but that fact could be a coincidence. The same color match could be obtained with innumerable other materials available to a forger. The first rule of good forgery has always been to use authentic materials. A forger otherwise good enough to have faked a Vinland Map should have obtained his color from an authentic material like tannin instead of an exotic titanium dioxide mixture. Therefore, one is moved to investigate the possibility of another hypothesis. Might it be possible that the map was originally titanium-free but the ink later became contaminated somehow with modern anatase titanium dioxide?

Paleographers maintain that sometime in its recent past the document has been washed or cleaned with a chemical. That hypothesis is corroborated by the British Museum's microscopical examination of the parchment's wormholes. The examination focused on the wormhole lining that bookworms always leave. In this case the lining has apparently been removed by the action of some chemical agent.[6]

A traditional way of cleaning documents was by bleaching. Nowadays conservators would be aghast at the idea of bleaching a parchment manuscript, but in the 1950s, when the Vinland Map was putatively still in private hands, it was common to "spruce up" antiquities to increase their sale value. The fact that the wormholes were patched shows that the owner held appearance above historical value. The bible of conservators at that time was the 1937 edition of Plenderleith, which advocated the same treatment for parchment manuscripts as for paper: regular household bleaching fluid,* sodium hypochlorite.[7]

A hypothesis that the map was bleached is consistent with the appreciable elemental percentage of sodium in the analysis. Traces of sodium were found in the plain, uninked parchment areas and larger amounts of sodium were found in the ink itself. Sodium has no function in any known ink recipe nor, in such quantity, in any white pigments. Nor can it be accounted for as common salt, sodium chloride (NaCl), from perspiring fingers. Its ratio to the chlorine in the ink (2:1 by weight) combined with the fact that chlorine's atom is half again as heavy as sodium's rules out its occurrence with the formula NaCl. However, sodium hypochlorite bleach, NaOCl, as it decomposes releases gaseous chlorine, leaving the sodium free to combine with atmospheric CO_2 and H_2O and then to appear in the observed ratio to other elements.[8]

Plenderleith described a bleaching method to be used on a document whose ink was unknown and possibly fugitive.[9] The method avoids the more usual procedure of immersion in bleaching fluid. Instead, a piece of dry tissue paper is laid on the face of the document. Then one brushes liquid bleach onto this paper and lets it soak through, peeling away the tissue before drying occurs. Now, it has been asserted that an unidentified private family library was the 1950s provenance of the Vinland Map,[10] and the kind of tissue paper a private family library would have on hand would be standard typewriter tissue. Its size would have been just perfect for insertion into this map's folio as it is bound with the Tartar Relation. However, in the 1940s and 1950s, some high-grade tissues, particularly onionskins and bible papers, were opacified with thin white coatings and fillers comprising exactly the pigments that were found in the ink.[11]

The coating or filler was held to the tissue by a binder of starch or casein. These are poorly soluble in water but are readily alkali-soluble. The alkaline pH level of commercial hypochlorite bleach would soften and loosen the binder of the paper pigments.[12] If the binder of the map ink were also alkali-soluble, then the viscid, pigment-laden paper coating would be in intimate contact with the viscid, pigment-free map ink binder. The slightest mechanical agitation, as from a brush applying the bleach, would mix the pigments into the ink and even under the edges of the carbon flakes. Transfer would be enhanced by the washing action of the advancing wet front as well as by gravity. When the two vehicles were separated and dried, the particles that entered the ink binder would be retained and others on the bare parchment perhaps not.

This author viewed the Vinland Map at Yale in 1996 after not having seen the original since the 1970s. My impression was that the yellow-brown ink seemed even paler than it had then. Others have expressed similar opinions. Perhaps there are residues from the hypothetical cleaning that are still continuing their effect. Yale University permitted this author to have a sample of the ink binder for solubility

---

*More recent editions of Plenderleith endorse the modern practice of never bleaching parchment.

testing. Nondestructive solubility testing of a microscopic sample is not easy, and my results need to be confirmed by others. However, it did appear that exposing the particle to the vapors of an alkaline solvent made the particle's sharp edges become rounded. And when the particle was subjected to immersion in a moderately strong alkali, it gradually wasted away.

In order to investigate the soundness of this transfer hypothesis, I have performed another experiment. I subjected a pure-tannin pseudo ink line on paper for half a minute to household bleach through tissue paper. For the tissue applique, I used pigmented paper from a Bible printed in 1952. It was necessary to use paper of that vintage, because more recently paper manufacturers have substituted insoluble synthetic binders for the casein used in the 1950s.

Under a 2000x immersion microscope, it was clear that the sample retained large quantities of pigment particles. The simulated ink exhibited numerous characteristics that are identical to features observed in the Vinland Map ink. First, as could be observed with lateral scanning under the microscope, pigment was retained by the ink binder but not by the bare paper beyond the edge of the ink line. Second, as could be observed by optical sectioning with varying focus, pigment particles penetrated throughout the interior as well as adhering to the surface of the ink binder.

The third point has to do with variability of pigment concentration. The Mc-Crones told me they found that subdivided ink particles from any one given sample point tended to agree with each other and indicated homogeneity within the individual micro-sample.[13] However, they found very large inhomogeneities of pigment concentration from one micro-sample point to another point within the document, ranging from 3 percent to 45 percent. The variation cannot be attributed to the smallness of the samples, in view of the aforementioned uniformity among the still smaller subdivisions. Neither can it be attributed to separately mixed batches of forger's ink, unless a different batch were used at virtually every point. Nor could the inhomogeneity be explained as a poorly mixed single batch, for then the draftsman would have had to dip his pen into a different part of the mix for virtually every point.

Notably, the simulated ink likewise had highly variable amounts of pigment from one area to another, but within definite micro-areas the amount was fairly constant. In the case of the simulated ink the cause of this phenomenon is clearly the fact of irregularity of transfer between the appliqué and the specimen surface. The map ink's variability could be similarly explained.* While it is clear that there is more to be

---

*Anyone wishing to repeat this experiment may be guided by some points of my own experience. I conducted the transfer process on delimed parchment and on paper. Deliming was necessary in order to avoid having the titanium pigments be optically camouflaged by the lime pigments. Under circumstances that I thought were all the same, I got results that included no transfer, transfer to ink only, and transfer to ink and parchment. The controlling parameters seemed to be alkalinity, time, and agitation, but these factors interact in a complicated way. The alkalinity increases as the bleach dries out, and it is important that the agitation be applied when the pH is just right to dissolve both the tissue binder and the ink binder. This would further account for the variability of pigment in the map. In addition, the specific Vinland Map parchment sheet was irregularly limed and cured, creating local pH variations.

learned about the details of transfer, there is no doubt whatsoever that this transfer mechanism does work.

Therefore, the McCrone conclusion of forgery is not a necessary, inescapable conclusion. The alternative hypothesis, of a misguided attempt in the 1950s to clear up a dirty old map to enhance its sales potential, is equally as plausible. The map's authenticity or inauthenticity cannot be established on the basis of the anachronistic pigments, and some other means must be used by those who feel a need to know with certainty.

NOTES

CHAPTER I INTRODUCTION

1. Anne Stein Ingstad, *Excavation of a Norse Settlement at L'Anse aux Meadows, Newfoundland, 1961–1968,* vol. 1 of *The Norse Discovery of America* (Oslo: Norwegian Univ. Press, 1985); Helge Marcus Ingstad, *The Historical Background of the Norse Settlement Discovered in Newfoundland,* vol. 2 of *The Norse Discovery of America.*

2. David Woodward, "Medieval *Mappaemundi,*" in J. B. Harley and David Woodward, eds., *The History of Cartography,* vol. 1 (Chicago: Univ. of Chicago Press, 1987), 318–21.

3. Jean Piaget, *Genetic Epistemology* (New York, 1970); idem, *Psychology and Epistemology* (New York, 1971); Michel Foucault, *The Archaeology of Knowledge* (New York, 1972).

4. Gunther S. Stent, "Prematurity and Uniqueness in Scientific Discovery," *Scientific American* (Dec. 1972), 93. Hans Furth, *Piaget and Knowledge* (Englewood Cliffs, N.J., 1969), 15.

5. Samuel Eliot Morison, *Admiral of the Ocean Sea* (Boston: Little, Brown, 1942), 92.

6. Charles Van Doren, *A History of Knowledge* (New York: Birch Lane, 1991), 174.

7. Carlos Sanz, "The Discovery of America," *Terrae Incognitae* (1974), 6:83–84.

8. Furth, *Piaget and Knowledge,* 201.

9. Harry Beilin and Peter B. Poufall, eds., *Piaget's Theory: Prospects and Possibilities* (Hillsdale, N.J.: L. Erlbaum, 1992).

10. James R. Enterline, *Viking America* (Garden City, N.Y.: Doubleday, 1972).

11. Morison, *Admiral of the Ocean Sea,* 68.

12. Pauline Moffitt Watts, "Prophecy and Discovery: On the Spiritual Origins of Christopher Columbus's 'Enterprise of the Indies,'" *American Historical Review* (Feb. 1985), 90:73–102.

13. For example, see John Dyson, *Columbus: For Gold, God and Glory* (New York: Simon & Schuster, 1991), 15–19, 200–213; Ian Wilson, *The Columbus Myth* (London: Simon & Schuster, 1991), chap. 6.

14. Thomas S. Kuhn, *The Structure of Scientific Revolutions* (Chicago: Univ. of Chicago Press, 1962), 86.

15. Stephen Greenblatt, ed., *New World Encounters* (Berkeley: Univ. of California Press, 1993).

16. Louis DeVorsey, "Silent Witnesses: Native American Maps," *The Georgia Review* (Winter 1992), 46 (4): 709–26.

17. G. Malcolm Lewis, "Misinterpretation of Amerindian Information as a Source of Error on European American Maps," *Annals of the Association of American Geographers* (Washington, D.C., 1987), 77:542–63.

18. Enterline, *Viking America.*

19. Gwyn Jones, *A History of the Vikings* (New York: Oxford Univ. Press, 1968), 311; Fridtjof Nansen, *In Northern Mists,* 2 vols. (London: Heinemann, 1911), 2:115; Helge Ingstad, *Land under the Pole Star* (New York: St. Martin's, 1966), 316–17.

20. Jette Arneborg, "Contact between Eskimos and Norsemen in Greenland," *Beretning fra tolvte tværfaglige vikingesymnposium, Aarhus Universitet* (Højbjerg: Forlaget Hikuin of Afdeling for Middelalder-arkæologi, 1993), 23–35.

21. Therkel Mathaissen, "Inugsuk, A Medieval Eskimo Settlement," *Meddelelser om Grönland* (Copenhagen: Reitzel, 1931), 77:284 ff; Erik Holtved, "Archaeological Investigations in the Thule District," *Meddelelser om Grönland,* vol. 141 (Copenhagen: Reitzel, 1944); Junius Bird, "Archaeology in the Hopedale Area, Labrador," *American Museum of Natural History, Anthropological Papers* (New York, 1945), 39:179–81.

22. Richard H. Jordan, "Neo-Eskimo Prehistory of Greenland," *Handbook of North American Indians,* vol. 5, *Arctic* (Washington, D.C.: Smithsonian Institution, 1984), 543–45.

23. Eric Alden Smith in Robert McGhee, "Disease and the Development of Inuit Culture," *Current Anthropology* (Dec. 1994), 35 (5): 584; Robert McGhee, "Contact between Native North Americans and the Medieval Norse: A Review of the Evidence," *American Antiquity* (1984), 49 (1): 4–26; Robert McGhee and Magnus Einarsson, "Greenlandic Eskimos and Norse: A Parallel Tradition from Greenland and Iceland?" *Folk* (Copenhagen, 1983), 25:58.

24. Knut Bergsland, "De norrøne lånord i grønlandsk," *Maal og Minne* (Oslo, 1986), 55–65.

25. Jette Arneborg, "Exchanges between Norsemen and Eskimos in Greenland," *Cultural and Social Research in Greenland 95/96,* ed. Brigitte Jacobsen (Nuuk, Greenland: Ilisimatusarfik, 1996), 11–21.

26. Franz Boas, *The Central Eskimo* (Washington, D.C., 1888; Lincoln: Univ. of Nebraska Press, 1964), 235.

27. John Spink and D. W. Moodie, "Eskimo Maps from the Canadian Eastern Arctic," *Cartographica,* vol. 5 (Toronto: Univ. of Toronto Press, 1972); John MacDonald, *The Arctic Sky: Inuit Astronomy, Star Lore, and Legend* (Toronto: Royal Ottawa Museum, 1998), chap. 6; Mark Warhus, *Another America: Native American Maps and the History of Our Land* (New York: St. Martin's Press, 1997), 126–37.

28. Renee Fossett, "Mapping Inuktitut: Inuit View of the Real World," in *Reading beyond Words: Contexts for Native History,* ed. Jennifer S. H. Brown and Elizabeth Vilbert (Peterborough, Ont.: Broadview Press, 1996), 74–94.

29. P. D. A. Harvey, *The History of Topographical Maps: Symbols, Pictures and Surveys* (London: Thames & Hudson, 1980), 37; *Gustav Holm Samlingen* (Grönlandssekretariatet pá Nationalmuseet), 47.

30. G. Malcolm Lewis, "Maps, Mapmaking, and Map Use by Native North Americans-Arctic," in *The History of Cartography,* vol. 2, ed. David Woodward and G. Malcolm Lewis, (Chicago: Univ. of Chicago Press, 1998), book 3:154–70; Hugh Brody, *The Other Side of Eden* (New York, 2000), 293 ff.

31. Woodward and Lewis, *The History of Cartography,* vol. 2, book 3, p. 3.

32. Edmund Carpenter, *Eskimo Realities* (New York: Holt Rinehart Winston, 1973), 27; Brody, *The Other Side of Eden,* 33–34, 51.

33. Robert A. Rundstrom, "Maps, Man, and Land in the Cultural Cartography of the Eskimo (Inuit)" (Ph.D. diss., University of Kansas, 1987; University Microfilms Pub. No. AAC8813443; idem, "A Cultural Interpretation of Inuit Map Accuracy," *Geographical Review* (New York: American Geographical Society, 1990), 80:155–68.

34. P. D. A. Harvey, "Local and Regional Cartography in Medieval Europe," in Harley and Woodward, *History of Cartography,* vol. 1, 464 ff.; idem, *Medieval Maps* (London: British Library, 1991).

35. James R. Brown, *Smoke and Mirrors: How Science Reflects Reality* (London: Routledge, 1993), 4.

36. Rudolf Simek, *Altnordische Kosmographie* (Berlin: Walter de Gruyter, 1990), 58 ff, 317 ff.

37. Heinrich Winter, "The Changing Face of Scandinavia and the Baltic in Cartography up to 1532," *Imago Mundi* (Leiden, 1955), 12:45–54.

38. Robert H. Fuson, *Legendary Islands of the Ocean Sea* (Sarasota, Fla.: Pineapple Press, 1995), 161–64.

39. Furth, *Piaget and Knowledge,* 15

40. T. G. R. Bower, "The Object in the World of the Infant," *Scientific American* (Oct. 1971), 37; idem, "Repetitive Processes in Child Development," *Scientific American* (Nov. 1976), 45.

41. James A. Bell, *Reconstructing Prehistory* (Philadelphia: Temple Univ. Press, 1994), chap. 5.

42. See illustration in W. G. L. Randles, *De la terre plate au globe terrestre* (Paris: Libraire Armand Colin, 1980), 30.

43. Randles, *De la terre plate,* chap. 1.

44. Kuhn, *Structure of Scientific Revolutions,* chap. 4, and pp. 62–65.

45. Nansen, *Northern Mists,* 1:18.

46. Ibid., 88, 98.

47. Ibid., 15–19.

48. Takashi Irimoto and Takako Yamada, eds., *Circumpolar Religion and Ecology* (Tokyo: Univ. of Tokyo Press, 1994), part 3.

49. MacDonald, *Arctic Sky,* 107 ff.

50. Douglas D. Anderson, "Prehistory of North Alaska," *Handbook of North American Indians,* vol. 5, *Arctic* (Washington, D.C.: Smithsonian Institution, 1984), 91.

51. MacDonald, *Arctic Sky.*

52. Ibid., 201, 114.

53. Ibid., 118.

54. Usama Fayyad, Gregory Piatetsky-Shapiro, and Padhraic Smyth, "The KDD Process for Extracting Useful Knowledge from Volumes of Data," *Communications of the ACM* (Nov. 1996), 39 (11): 29.

55. See review of Charles Hapgood, *Maps of the Ancient Sea Kings* in *Geographical Journal* (London, 1967), 133:394–95.

56. Paul Gallez, *La cola del dragon* (Bahia Blanca, Argentina: Instituto Patagonico, 1990).

57. James Enterline, "The Southern Continent and the False Strait of Magellan," *Imago Mundi* (Amsterdam, 1972), 26:48–58; Benjamin B. Olshin, "A Sea Discovered: Pre-Columbian Conceptions and Depictions of the Atlantic Ocean" (Ph.D. diss., University of Toronto, 1993), chap. 10.

58. J. H. Parry, "Old Maps Are Slippery Witnesses," *Harvard Magazine* (Apr. 1976), 78 (8): 32–41.

59. Benoit B. Mandelbrot, "How Long Is the Coastline of Britain?" *Science* (Washing-

ton, D.C., May 5, 1967), 156:636; idem, *The Fractal Geometry of Nature* (San Francisco: W. H. Freeman, 1982), 34.

60. Benoit B. Mandelbrot, "Stochastic Models for the Earth's Relief, the Shape and the Fractal Dimension of the Coastlines, and the Number-Area Rule for Islands," *Proceedings of the National Academy of Sciences, U.S.A.* (Washington, D.C., Oct. 1975), 72 (10): 3826.

61. M. J. Blakemore and J. B. Harley, "Concepts in the History of Cartography," *Cartographica* (Monograph No. 26) (Toronto: Winter 1980), 17 (4): 75.

62. Mandelbrot, *Fractal Geometry,* 22.

63. Mark Monmonier, *Drawing the Line: Tales of Maps and Cartocontroversy,* (New York: Henry Holt, 1995), 148–69.

64. Corey S. Powell, "Science and the Citizen," *Scientific American* (Sept. 1994), 14.

65. Brown, *Smoke and Mirrors,* 13–15.

66. Ian W. D. Dalziel, "Earth before Pangea," *Scientific American* (Jan. 1995), 58–63.

CHAPTER 2 CLAUDIUS CLAVUS

1. Wilcomb E. Washburn, "Representation of Unknown Lands in Fourteenth-, Fifteenth-, and Sixteenth-Century Cartography," *Agrupamento de Estudos de Cartografia Antiga* (Coimbra, 1969), 35:4–6.

2. Gustav Storm, "Den danske geograf Claudius Calvus eller Nicolaus Niger," *Ymer,* parts 1 & 2 (Stockholm, 1889), 9:129–46; parts 3–5 (Stockholm, 1891), 11:13–37. Coordinate table reproduced in 11:24–38.

3. Joseph Fischer, "Claudius Clavus," *Historical Records and Studies, U.S. Catholic Historical Society* (New York, 1911), part 1, vol. 6.

4. Fridtjof Nansen, *In Northern Mists,* 2 vols. (London: Heinemann, 1911), 2:260–61.

5. Andrew Fossum, *The Norse Discovery of America* (Minneapolis, 1918), 55–65.

6. Fischer, "Claudius Clavus."

7. Knut Bergsland, "De norrøne lånord i Grønlandsk," *Maal og Minne* (Oslo, 1986), 55–65.

8. J. Brian Harley, "Deconstructing the Map," *Cartographica,* 26 (2): 1–20.

9. Regarding early Thule sites on Bering Sea, see Don E. Dumond, "Prehistory of the Bering Sea Region," *Handbook of North American Indians,* vol. 5, *Arctic* (Washington, D.C.: Smithsonian Institution, 1984), 103–4.

10. William Edward Parry, *Journal of a Second Voyage for the Discovery of a Northwest Passage* (New York: Greenwood Press, 1969), 198.

11. Frederick William Beechey, *Narrative of a Voyage to the Pacific and Beering's Strait* (London: H. Colburn & R. Bentley, 1831), 1:290–91, 292.

12. Steven Davis and Dorothy F. Prescott, "Symbolism of Traditional Aboriginal Geographical Knowledge of Australia," *Globe* (journal of the Australian Map Circle, Inc.) (Melbourne, 1998), 46:1–20; Daniel Boorstin, "The Lost Arts of Memory," *The Discoverers* (New York: Random House, 1983), chap. 60.

13. Beechey, *Narrative of a Voyage,* 2:534.

14. Guy Mary-Rousselière, *Qitdlarssuaq: The Story of a Polar Migration* (Winnipeg: Wuertz, 1991).

15. Robert McGhee, "Thule Prehistory of Canada," *Handbook of North American Indians,* vol. 5, *Arctic* (Washington, D.C.: Smithsonian Institution, 1984), 368–76.

16. Susan Rowley, "The Significance of Migration for the Understanding of Inuit Cultural Development in the Canadian Arctic" (Ph.D. diss., University of Cambridge, 1985), 194.

17. Karen M. McCullough, *The Ruin Islanders,* Archaeological Survey of Canada, Mercury Series Paper 141 (Hull, Que.: Canadian Museum of Civilization, 1989), 258.

18. Ibid., 303.

19. W. R. Tobler, *Bidimensional Regression: A Computer Program* (Santa Barbara: Univ. of California Geography Dept., 1977).

20. Robert McGhee, "Disease and the Development of Inuit Culture," *Current Anthropology* (1994), 35 (5): 590; Douglas D. Anderson, "Prehistory of North Alaska," *Handbook of North American Indians,* vol. 5, *Arctic* (Washington, D.C.: Smithsonian Institution, 1984), 92.

21. Axel A. Björnbo and Carl S. Petersen, *Der Däne Claudius Clausson Swart* (Innsbruck: Wagner, 1909), 132–52; 235–42.

22. Ibid.

23. Cecil Brown, "Where Do Cardinal Directions Terms Come From?" *Anthropological Linguistics* (1983), 25:121–61.

24. John MacDonald, *The Arctic Sky* (Toronto: Royal Ottawa Museum, 1998), 173–4.

25. Michael Fortescue, "Eskimo Orientation Systems," *Meddelelser om Grönland, Man and Society* (Copenhagen: Commission for Scientific Research in Greenland, 1988), 11:10 map 2b, 15–16, 20–25.

26. Robert McGhee, "Contact between Native North Americans and the Medieval Norse: A Review of the Evidence," *American Antiquity* (1984), 49 (1): 22; James R. Enterline, *Viking America* (Garden City, N.Y.: Doubleday, 1972), chap. 7.

27. Washburn, "Representation of Unknown Lands," 9.

28. Vilhjalmur Stefansson, *Ultima Thule* (New York: Macmillan, 1940), chap. 1.

29. Björnbo and Petersen, *Der Däne Claudius,* 96.

CHAPTER 3 THE *INVENTIO FORTUNATAE* AND MARTIN BEHAIM

1. B. F. De Costa, "Arctic Exploration," *American Geographical Society Journal* (New York, 1880), 12:161–63.

2. E. G. R. Taylor, "A Letter Dated 1577 from Mercator to John Dee," *Imago Mundi* (Berlin, 1956), 13:56–68.

3. Ibid., 65.

4. Delno West, "*Inventio Fortunata,* Medieval Metaphor, and Arctic Cartography," presented at Society for the History of Discoveries annual meeting, Mackinac Island, Mich., 1994; idem, "Inventio fortunata and Polar Cartography 1360–1700" (paper presented at the conference "De-Centering the Renaissance: Canada and Europe in Multi-Disciplinary Perspective 1350–1700," Victoria University in the University of Toronto, March 7–10, 1996).

5. Taylor, "A Letter Dated 1577," 63.

6. E. G. Ravenstein, *Martin Behaim, His Life and His Globe* (London: G. Philip & Son, 1908), 71.

7. Tryggvie Oleson, *Early Voyages* (Toronto: McClelland & Stewart, 1963), 98–99.

8. Taylor, "A Letter Dated 1577," 56–68.

9. George H. T. Kimble, *Geography of the Northlands* (New York: American Geographical Society and J. Wiley, 1955), 363, fig. 18.

10. David F. Parmelee et al., "The Birds of Southeastern Victoria Island and Adjacent Small Islands," National Museum of Canada Bulletin No. 222, Biological Series No. 78 (Ottawa, 1967), 68; L. L. Snyder, *Arctic Birds of Canada* (Toronto: Univ. of Toronto Press, 1957), 105.

11. United States Hydrographic Office, *Sailing Directions for Northern Canada: The Coast*

of Labrador Northward of St. Lewis Sound, the Northern Coast of the Canadian Mainland, and the Canadian Archipelago (Washington, D.C.: Government Printing Office, 1951–65), 357.

12. Chet Van Duzer, "The Mythology of the Northern Polar Regions: Inventio fortunata and Buddhist Cosmology," *At the Edge: Exploring New Interpretations of Past and Place in Archaeology, Folklore, and Mythology* (Mar. 1998), 9:8–16.

13. U.S. Hydrographic Office, *Sailing Directions,* 229–31; Canadian Hydrographic Service, *Labrador and Hudson Bay Pilot,* 2nd ed. (Ottawa, 1965), 305.

14. U.S. Hydrographic Office, *Sailing Directions,* 303–8.

CHAPTER 4 THE YALE VINLAND MAP

1. Helen Wallis et al., "The Strange Case of the Vinland Map," *Geographical Journal* (London, 1974), 140:212 ff.

2. R. A. Skelton, Thomas E. Marston, and George D. Painter, *The Vinland Map and the Tartar Relation* (New Haven: Yale Univ. Press, 1965, 1995), 3–16.

3. T. A. Cahill et al., "The Vinland Map, Revisited: New Compositional Evidence on Its Inks and Parchment," *Analytical Chemistry* (1987), 59:829 ff.

4. James Enterline, "A Transport Mechanism for Accidental Pigments in an Ink Line," *Seventh International Conference on the History of Cartography* (proceedings) (Washington, D.C., 1977).

5. Kirsten A. Seaver, "The 'Vinland Map': Who Made It and Why? New Light on an Old Controversy," *Map Collector* (Tring, Herefordshire: Spring 1995), 70:32–40; idem, "The Mystery of the 'Vinland Map' Manuscript Volume," *Map Collector* (Tring, Herefordshire: Spring 1996), 24–29. idem, "The Vinland Map," *Mercator's World* (Mar.–Apr. 1997), 42–47.

6. Thomas A. Cahill and Bruce H. Kusko, "Compositional and Structural Studies of the Vinland Map and Tartar Relation," in Skelton, Marston, and Painter, *The Vinland Map* (1995), xxix–xxxix.

7. "Crime Lab," episode in Discover Magazine series, aired on Discovery Channel, Apr. 28, 1996.

8. Alf Mongé and O. Landsverk, *Norse Medieval Cryptography in Runic Carvings* (Glendale, Calif.: Norseman Press, 1967), 118–22; James Enterline, "Cryptography in the Yale Vinland Map," *Terrae Incognitae* (1991 plus errata sheet 1992), 23:13–27; Louis Kruh, "Book Reviews," *Cryptogram* (Wilbraham, Mass.: American Cryptogram Assn., 1993), 59 (1): 15; idem, "Vineland Cryptography," *Cryptologia* (Terre Haute, Ind.: Rose-Hulman Institute, 1993), 17 (1): 107–8.

9. Paul Saenger, "Vinland Re-read," *Imago Mundi* (1998), 50:199–202.

10. Skelton, Marston, and Painter, *Vinland Map,* vii–lx.

11. Ibid., 110, 114.

12. Erik Wahlgren, *The Vikings and America* (New York: Thames & Hudson, 1986), 143.

13. *Proceedings of the Eighth Viking Conference (1977)* (Odense, 1981), 3–8.

14. Helge Marcus Ingstad, *The Historical Background of the Norse Settlement Discovered in Newfoundland,* vol. 2 of *The Norse Discovery of America* (Oslo: Norwegian Univ. Press, 1985), 307–13.

15. James R. Enterline, *Viking America* (Garden City, N.Y.: Doubleday, 1972), 36–40.

16. "Discovery at Canadian Museum of Civilization Suggests the Norse Visited Baffin Island in the Middle Ages," http://www.civilization.ca/cmc/cmceng/prio8eng.html, 1999.

17. Samuel Eliot Morison, *The European Discovery of America: The Northern Voyages* (New York: Oxford Univ. Press, 1971), 69.

18. Franz Boas, *The Central Eskimo* (Washington, D.C., 1888; Lincoln, Neb., 1964), 235.

19. E. G. R. Taylor, "The Vinland Map," *Journal of Navigation* (London: John Murray Publishers for Royal Institute of Navigation, 1974), 27:195–205.

20. Axel A. Björnbo, "Cartographia Groenlandica," *Meddelelser om Grönland* (Copenhagen, 1912), 48:125.

21. Helen Wallis, "The Vinland Map: Fake, Forgery, or *jeu d'esprit*," *Map Collector* (Tring, 1990), 53:2–6.

CHAPTER 5   INTRODUCTION TO THE CHRONOLOGICAL SURVEY

1. Henry Harrisse, *The Discovery of North America* (London, 1892; Amsterdam: N. Israel, 1961).

2. R. A. Skelton, *Maps: A Historical Survey of Their Study and Collecting* (Chicago: Univ. of Chicago Press, 1972), 28–33.

3. Michael C. Andrews, "Study and Classification of Medieval Mappaemundi," *Archaeologia* (Oxford, 1926), 75:65.

4. J. B. Harley and David Woodward, eds., *The History of Cartography,* vol. 1 (Chicago: Univ. of Chicago Press, 1987), 334–42.

5. Rudolf Simek, *Altnordische Kosmographie* (Berlin: Walter de Gruyter, 1990), fig., p. 106; idem, *Heaven and Earth in the Middle Ages: The Physical World before Columbus,* trans. Angela Hall (Rochester, N.Y.: Boydell Press, 1996), 37–38, 40–43.

6. Andrews, *Study and Classification,* 69.

7. W. G. L. Randles, *De la terre plate au globe terrestre* (Paris: Libraire Armand Colin, 1980), 15–17.

8. Andrews, *Study and Classification,* 67.

9. B. L. Gordon, "Sacred Directions, Orientations and the Top of the Map," *History of Religions* (Chicago: Univ. of Chicago Press, 1971), 10:211–27.

10. Charles O. Frake, "Dials: Representation of Cognitive Systems," in *The Ancient Mind: Elements of Cognitive Archaeology,* ed. Colin Renfrew (Cambridge: Cambridge Univ. Press, 1994), 127; G. J. Marcus, *The Conquest of the North Atlantic* (New York: Oxford Univ. Press, 1981), chap. 15.

11. Thorkild Ramskou, *Vikingernes Hverdag / Everyday Viking Life* (Danish/English parallel texts) (Copenhagen: Rhodos, 1967), 142–44; C. L. Vebaek, "Ten Years of Topographical and Archaeological Investigations in the Medieval Norse Settlements in Greenland," *Proceedings of the 32nd International Congress of Americanists* (Copenhagen: Munksgaard, 1956), 732–43; Kåre Prytz, *Westward before Columbus* (Weston, Conn.: Norumbega Books, 1991), 8.

12. James Enterline, ed., "Sun Ray Disk White Paper," http://www1.minn.net/keithp/vdisk.htm, 1999.

13. L. Bagrow and R. A. Skelton, *History of Cartography* (Cambridge, 1964; Chicago: Precedent Publishing, 1985), 62; Lloyd Brown, *The Story of Maps* (1949; reprint, New York: Dover Publications, 1979), 126–27; R. A. Skelton, Thomas E. Marston, and George D. Painter, *The Vinland Map and the Tartar Relation* (New Haven: Yale Univ. Press, 1965, 1995), 169.

14. C. R. Beazley, *The Dawn of Modern Geography,* 3 vols. (Oxford, 1896–1906; reprint, New York, 1949), 3:508–11.

15. L. Hongre, G. Hulot, and A. Khokhlov, "An Analysis of the Geomagnetic Field over the Past 2000 Years," *Physics of the Earth and Planetary Interiors* (Amsterdam: North Holland, 1998), 106:311–35, esp. fig. 7a.

16. Tony Campbell, "Portolan Charts from the Late Thirteenth Century to 1500," in Harley and Woodward, *History of Cartography,* 1:441.

CHAPTER 6 EARLY SCANDINAVIAN GEOGRAPHY

1. Edward L. Stevenson, *The Geography of Claudius Ptolemy* (New York, 1932); Lloyd Brown, *The Story of Maps* (1949; reprint, New York: Dover Publications, 1979), chap. 3.

2. Howard R. Patch, *The Other World* (Cambridge, 1950); Fridtjof Nansen, *In Northern Mists,* 2 vols. (London: Heinemann, 1911), 1:345–53; John Thacher, *Christopher Columbus* (New York, 1903), 1:493–511; Philip O. Spann, "Sallust, Plutarch, and the 'Isles of the Blest,'" *Terrae Incognitae* (1977), 9:75–80.

3. Gwyn Jones, *The Norse Atlantic Saga* (New York: Oxford Univ. Press, 1964); C. R. Beazley, *The Dawn of Modern Geography,* 3 vols. (Oxford, 1896–1906; reprint, New York, 1949), 2:17–28; Jón Jóhannesson, *A History of the Old Icelandic Commonwealth,* trans. Haraldur Bessanson (Winnipeg: Univ. of Manitoba Press, 1974).

4. Ibn Khordâdhbeh, *Le livre des routes et des provinces,* trans. C. Barbiea de Maynard (Paris, 1865); Nansen, *Northern Mists,* 2:196–97; Eric Oxenstierna, "The Vikings," *Scientific American* (May 1967).

5. Niels Lund, ed., *Two voyagers at the Court of King Alfred: The Ventures of Othere and Wulfstan, Together with the Description of Northern Europe from the Old English Orosius* (York: William Sessons Ltd., 1984); Joseph Bosworth, *A Description of Europe, Africa, etc., by Alfred the Great* (London, 1855); Nansen, *Northern Mists,* 1:170–80.

6. Magnus Magnusson and Herman Palsson, *The Vinland Sagas* (New York: New York Univ. Press, 1966); James R. Enterline, *Viking America* (Garden City, N.Y.: Doubleday, 1972); Jones, *Norse Atlantic Saga;* Vilhjalmur Stefansson, *Greenland* (Garden City, N.Y.: Doubleday, 1942); Daniel Brunn, "The Icelandic Colonization of Greenland," *Meddelelser om Grönland,* vol. 57 (Copenhagen, 1918); Aage Roussell, "Farms and Churches in the Medieval Norse Settlements of Greenland," *Meddelelser om Grönland,* vol. 89 (Copenhagen, 1941).

7. Konrad Miller, *Mappaemundi: Die ältesten Weltkarten,* 6 vols. (Stuttgart, 1895), 3:29–37; Beazley, *Dawn of Modern Geography,* 2:559–63, 608–12.

8. Francis J. Tschan, *Adam of Bremen's History of the Archbishops of Hamburg-Bremen* (New York: Columbia Univ. Press, 1959), 218–19; Axel A. Björnbo, "Cartographia Groenlandica," *Meddelelser om Grönland* (Copenhagen, 1912), 48:70; Beazley, *Dawn of Modern Geography,* 2:514–48.

9. Enterline, *Viking America,* chap. 3; Kevin P. Smith, "Re: 'vin' vs. 'viin'" in ONN (Old Norse Net) at onn@hum.gu.se (July 26, 1998).

10. R. A. Skelton, Thomas E. Marston, and George D. Painter, *The Vinland Map and the Tartar Relation* (New Haven: Yale Univ. Press, 1965, 1995), 223–26; 255–61.

11. Nansen, *Northern Mists,* 2:1.

12. Halldor Harmannsson, "The Problem of Wineland the Good," *Islandica* (Ithaca, N.Y.: Cornell Univ. Press, 1936), 25:3.

13. Rudolf Simek, *Altnordische Kosmographie* (Berlin: Walter de Gruyter, 1990), 102 ff.

14. Nansen, *Northern Mists,* 2:2; Rudolf Simek, "Elusive Elysia," *Sagnaskemmtun* (Vienna: Hermann Böhlaus, 1986), 247–75.

15. Nansen, *Northern Mists,* 2:1 n. 1.

16. Oliver Elton, "The Nine Books of the Danish History of Saxo Grammaticus," *Anglo-Saxon Classics,* 15 vols. (London: Norroena Society, 1907), 1:88.

17. Miller, *Mapae Mundi,* 3:21–29; Beazley, *Dawn of Modern Geography,* 2:563–66; 614–

17; Patrick Gautier Dalche, *La "Descriptio mappe mundi" de Hughes de Saint-Victor* (Paris: Etudes Augustiniennes, 1988), 183 n. 13.

CHAPTER 7 COMMUNICATION LINKS WITH GREENLAND

1. Fridtjof Nansen, *In Northern Mists,* 2 vols. (London: Heinemann, 1911), 2:208–9.

2. L. M. Larson, *The King's Mirror* (New York, 1917).

3. Vincent Cassidy, "The Location of Ginnungagap," in *Scandinavian Studies: Essays Presented to Henry Goddard Leach,* ed. Carl F. Bayerschmidt and Erik J. Friis (Seattle, 1965), 27–38.

4. F. X. Dillmann, "Ginnungagap," in *Reallexikon der Germanischen Altertumskunde,* ed. Johannes Hoops (Berlin, 1998), 12:118–23.

5. Nansen, *Northern Mists,* 2:34.

6. Halldor Hermannsson, *The Vinland Sagas* (Ithaca, N.Y., 1944); Magnus Magnusson and Herman Palsson, *The Vinland Sagas* (New York, 1966); Peter Hallberg, *The Icelandic Sagas* (Lincoln, Neb., 1962); Theodore Andersson, *The Problem of Icelandic Saga Origins* (New Haven, 1964).

7. Jesse L. Byock, "Egil's Bones," *Scientific American* (Jan. 1995), 82–87.

8. Jônas Kristjansson, "The Roots of the Sagas," *Sagnaskemmtun* (Vienna: Hermann Böhlaus, 1986), 183–200.

9. Gustav Storm, *Monumenta Historiaca Norvegiae* (Christiana, 1880); Halvdan Koht, *Den Eldste Noregs-Historia* (Oslo, 1950), 11–13; Nansen, *Northern Mists,* 1:255.

10. Robert McGhee, "Radiocarbon Dating and the Timing of the Thule Migration," in *Identities and Cultural Contacts in the Arctic,* ed. Martin Appelt, Joel Berglund, and Hans Christian Gullov (Copenhagen: Danish Polar Center Publication #8, 2000), 181–91.

11. Robert McGhee, "Contact between Native North Americans and the Medieval Norse: A Review of the Evidence," *American Antiquity* (1984), 22.

12. Kongelige Dansk videnskabernes selskab, *Grönlands Historiske Mondesmaerker* (Copenhagen, 1845), 3:238–43; Nansen, *Northern Mists,* 1:308–311.

13. Konrad Miller, *Mappaemundi: Die ältesten Weltkarten,* 6 vols. (Stuttgart, 1895), 3:37–43; C. R. Beazley, *The Dawn of Modern Geography* (Oxford, 1896–1906; reprint, New York, 1949), 2:568–69; 617–21.

14. P. D. A. Harvey, *Mappa Mundi: The Hereford World Map* (Toronto: Toronto Univ. Press, 1996); Miller, *Mappaemundi,* vol. 4; International Geographical Union, Commission on Early Maps, *Monumenta Cartographica Vetustioris Aevi* (Amsterdam, 1964), 1:197–202.

CHAPTER 8 THE UNSEEN BRIDGE

1. Arthur Middleton Reeves, *The Finding of Wineland the Good* (London, 1895; New York, 1967), 87–89; J. K. Tornöe, *Columbus in the Arctic?* (Oslo, 1965), 48–51.

2. F. S. Haydon, "Eulogium," *Rerum Britannicarum Medii Aevi Scriptores* (London, 1860), 9 (2): xxiv–xlvi, 78–79; Fridtjof Nansen, *In Northern Mists,* 2 vols. (London: Heinemann, 1911), 1:192, 2:31–32, 189–91; Diamond Jenness, "Eskimo String Figures," *Report of the Canadian Arctic Expedition* (Ottawa, 1924), 13 (B): 181.

3. G. J. Marcus, "The Greenland Trade Route," *Economic History Review* (London, 1954), ser. 2, 7:71–80; Reeves, *Finding of Wineland,* 162–64 n. 12; Vilhjalmur Stefansson, *Greenland* (Garden City, N.Y.: Doubleday, 1942), 91–94; 160–61; John A. Gade, *The Hanseatic Control of Norwegian Commerce during the Late Middle Ages* (Leiden, 1951), 45–46; Jón Jóhan-

nesson, *A History of the Old Icelandic Commonwealth,* trans. Haraldur Bessanson (Winnipeg: Univ. of Manitoba Press, 1974), 222–87.

4. Arthur Christopher Moule and Paul Pelliot, *Marco Polo: The Description of the World* 4 vols. (Paris, 1938–63); Ronald Latham, trans., *The Travels of Marco Polo* (London: Folio Society, 1958; New York: Abaris, 1982).

5. *Rymbegla* (Havniae, 1780).

6. Youssouf Kamal, *Monumenta Cartographica* (Cairo, 1936), tome 4, fasc. 1, fol. 1138; C. R. Beazley, *The Dawn of Modern Geography* (Oxford, 1896–1906; reprint, New York, 1949), 3:512–19; Edward L. Stevenson, *Portolan Charts, Their Origin and Characteristics* (New York, 1911); Tony Campbell, "Portolan Charts from the Late Thirteenth Century to 1500," in *The History of Cartography,* vol. 1, ed. J. B. Harley and David Woodward (Chicago: Univ. of Chicago Press, 1987).

7. Kamal, *Monumenta Cartographica,* tome 4, fasc. 1, fol. 1174; Konrad Kretchmer, "Marino Sanudo der ältere und die Karten des Petrus Vesconte," Gesellschaft für Erdkunde zu Berlin, *Zeitschrift* (Berlin, 1891), 26:352–70; Nansen, *Northern Mists,* 2:222–26.

8. Bernhard Degenhart and Annegrit Schmitt, "Marino Sanudo und Paolino Veneto," *Römisches Jahrbuch für Kunstgeschichte* (1973), 14:107.

9. Joachim Lelewel, *Géographie du Moyen Age* (Brussels, 1852–57), 2:25.

10. A. R. Hinks, "The Portolan Chart of Angellino Dalerto, 1325," *Reproductions of Early Manuscript Maps* (London, 1929), vol. 1; Nansen, *Northern Mists,* 2:226–30; William H. Babcock, *Early Norse Visits to North America* (Washington, D.C., 1913), 21–26.

11. Robert H. Fuson, *Legendary Islands of the Ocean Sea* (Sarasota, Fla.: Pineapple Press, 1995), 44–51.

12. Ibid., 46, 86.

13. William Thalbitzer, "Two Runic Stones," *Smithsonian Miscellaneous Collections* (Washington, D.C., 1952), vol. 116, no. 3; David Diringer, *The Alphabet* (New York, 1948), 507–24; Alf Mongé and O. G. Landsverk, *Norse Medieval Cryptography in Runic Carvings* (Glendale, Calif., 1967); A. Liestol, "Cryptograms in Runic Carvings: A Critical Analysis," *Minnesota History* (St. Paul, 1968), 41:34–42; Hans Karlgren, "Review of Mongé and Landsverk," *Scandinavian Studies* (Lawrence, Kans., 1968), 40:326–30; O. G. Landsverk, *Ancient Norse Messages on American Stones* (Glendale, Calif., 1969), chap. 4.

14. Nansen, *Northern Mists,* 2:100–103.

15. B. F. De Costa, *Sailing Directions of Henry Hudson* (Albany, 1869); Nansen, *Northern Mists,* 2:107–12.

16. Rachel Carson, *The Sea Around Us* (New York, 1963), 162–72; Seymour Tilson, "The Ocean," *International Science and Technology* (Feb. 1966), 36.

17. Kirsten A. Seaver, *The Frozen Echo* (Stanford, Calif.: Stanford Univ. Press, 1996), 119, 131.

18. Ibid., 108–12.

CHAPTER 9 LATE GREENLAND–BASED EXPLORATION

1. Arthur Middleton Reeves, *The Finding of Wineland the Good* (London, 1895; New York, 1967), 83; Fridtjof Nansen, *In Northern Mists,* 2 vols. (London: Heinemann, 1911), 2:36–38.

2. Theobald Fischer, *Raccolta di Mappamundi e carte nautice del XIII al XVI secolo* (Venice, 1881), vol. 5; idem, *Sammlung Mittelalterlicher Welt-und Seekarten* (Venice, 1886), 127–47; Nansen, *Northern Mists,* 2:234–36.

3. E. G. R. Taylor, "A Letter Dated 1577 from Mercator to John Dee," *Imago Mundi* (Amsterdam, 1956), 13:56−68; B. F. De Costa, "Arctic Exploration," *American Geographical Society Journal* (New York, 1880), 12:172−89.

4. Helge Marcus Ingstad, *The Historical Background of the Norse Settlement Discovered in Newfoundland,* vol. 2 of *Norse Discovery* (Oslo: Norwegian Univ. Press, 1985), 377−78, 392; Kirsten Seaver, *The Frozen Echo* (Stanford, Calif.: Stanford Univ. Press, 1996), 134.

5. R. A. Skelton, Thomas E. Marston, George D. Painter, *The Vinland Map and the Tartar Relation* (New Haven: Yale Univ. Press, 1965, 1995), 179 n. 138.

6. Helge Ingstad, *Land under the Pole Star* (New York: St. Martin's, 1966), 93.

7. Richard Henry Major, *The Voyages of the Venetian Brothers Nicolo and Antonio Zeno* (London, 1873); Frederick W. Lucas, *Annals of the Voyages of the Brothers Zeno* (London, 1898). Frederick Pohl, *Prince Henry Sinclair* (New York, 1974), chap. 11.

8. Robert H. Fuson, *Legendary Islands of the Ocean Sea* (Sarasota, Fla.: Pineapple Press, 1995), 38.

9. Marcel Destombes, "Fragments of Two Medieval World Maps at the Topkapu Sary Library," *Imago Mundi* (Amsterdam, 1955), 12:150−52; *Monumenta Cartographica Vetustioris Aevi* (Amsterdam, 1964), 1:203−5. G. R. Crone, *Maps and Their Makers* (London, 1953), chap. 3.

10. Nansen, *Northern Mists,* 2:112−13.

11. Michel Wolfe, "Norse Archaeology in Greenland since World War II," *American Scandinavian Review* (1961), 49:383.

12. Major, *Voyages of Nicolo and Antonio Zeno;* Lucas, *Annals of Voyages;* Pohl, *Prince Henry Sinclair,* chap. 11.

CHAPTER 10 FOUNDATIONS OF EUROPEAN MISUNDERSTANDINGS

1. Edward L. Stevenson, *The Geography of Claudius Ptolemy* (New York, 1932); Joseph Fischer, *Claudii Ptolemaei Geographiae,* 3 vols. (Turin, 1932).

2. Edmund J. P. Buron, ed., *Imago Mundi de Pierre d'Ailly,* 3 vols. (Paris: Librairie Orientale et Americaine, 1930); Edwin F. Keever, *Imago Mundi by Petrus Ailliacus translated from the Latin* (Wilmington, N.C., 1948).

3. Franz R. von Wieser, *Die Weltkarte des Albertin de Virga* (Innsbruck, 1912); *Monumenta Cartographica Vetustioris Aevi* (Amsterdam, 1964) 1:205−7.

4. Dana Durand, *The Vienna-Klosterneuburg Map Corpus* (Leiden: E. J. Brill, 1952), 213−15, and pl. 16.

5. G. Malcolm Lewis, "Native North American Cosmological Ideas and Geographical Awareness," in *North American Exploration,* ed. John L. Allen (Lincoln: Univ. of Nebraska Press, 1997), 1:98−99, 96−97, 121−22.

6. Robert H. Fuson, *Legendary Islands of the Ocean Sea* (Sarasota, Fla.: Pineapple Press, 1995), 49−50; E. G. R. Taylor, "A Letter Dated 1577 from Mercator to John Dee," *Imago Mundi* (Amsterdam, 1956), 60.

7. Helge Ingstad, *Land under the Pole Star* (New York: St. Martin's, 1966), 325−32.

8. Map Division, U.S. Library of Congress, Title Collection, World, 1422−39, *Ducier,* (Arlington, Va.); Robert F. Marx, *In Quest of the Great White Gods* (New York: Crown, 1992), 51−52, 306−26; William R. McGlone et al., *Ancient American Inscriptions* (Sutton, Mass.: Early Sites Research Society, 1993), 313−14.

9. Leo Bagrow, "Maps from the Home Archives of the Descendants of a Friend of Marco Polo," *Imago Mundi* (1948), 5:3.

10. Janice Poston, "The Rossi Collection," report, U.S. Library of Congress Geography and Map Division, July 5, 1992.

11. Plutarch, *The Lives of the Noble Grecians and Romans,* trans. John Dryden (New York: Modern Library, 1979), 683; Christoph F. Konrad, *Plutarch's Sertorius: A Historical Commentary* (Chapel Hill: Univ. of North Carolina Press, 1994).

12. R. A. Skelton, Thomas E. Marston, and George D. Painter, *The Vinland Map and the Tartar Relation* (New Haven: Yale Univ. Press, 1965, 1995), 158.

13. James E. Kelley, "Non-Mediterranean Influences that Shaped the Atlantic in the Early Portolan Charts," *Imago Mundi,* (1979), 31:27–33.

14. Fuson, *Legendary Islands,* chap. 11.

15. Armando Cortesao, "The North Atlantic Nautical Chart of 1424," *Imago Mundi* (Stockholm, 1953), 10:1–13; idem, *The Nautical Chart of 1424* (Coimbra, 1954); idem, "Pizzigano's Chart of 1424," *Revista da Universidade de Coimbra* (Coimbra, 1970), vol. 24; William Babcock, *Legendary Islands of the Atlantic* (New York, 1922), chap. 3; T. J. Westropp, "Brasil and the Legendary Islands of the North Atlantic," *Proceedings of the Royal Irish Academy* (1912), 30-C:223–60; Kåre Prytz, "The Antilia Chart of 1424 and Later Maps of America," *Norsk Geografisk Tidsskrift* (Oslo: Universitetsforlaget, 1977), 31:57–67; idem, *Westward before Columbus* (Weston, Conn.: Norumbega Books, 1991).

CHAPTER 11 NEWS PENETRATES THE ESTABLISHMENT

1. Joseph Fischer, "Claudius Clavus," *Historical Records and Studies, U.S. Catholic Historical Society* (New York, 1911), part 1, vol. 6; Axel A. Björnbo and Carl S. Petersen, *Der Däne Claudius Claussön Swart* (Innsbruck, 1909); Harald Sigurdsson, *Kortsaga Islands fra ondverou til loka 16 aldar* (Reykjavik, 1981), 259 ff.

2. Fischer, "Claudius Clavus," 85–101.

3. Dana Durand, *The Vienna-Klosterneuburg Map Corpus* (Leiden, 1952); Andrew R. Anderson, *Alexander's Gate, Gog and Magog, and the Enclosed Nations* (Cambridge, Mass., 1932), 72–74.

4. James R. Enterline, *Viking America* (Garden City, N.Y.: Doubleday, 1972), 40–41.

5. Andrew C. Gow, *The Red Jews: Antisemitism in an Apocapalyptic Age, 1200–1600* (Leiden: E. J. Brill, 1995), 66 ff.

6. R. A. Skelton, Thomas E. Marston, and George D. Painter, *The Vinland Map and the Tartar Relation* (New Haven: Yale Univ. Press, 1965, 1995), 124–27.

7. James Enterline, "Cryptography in the Yale Vinland Map," *Terrae Incognitae* (1991 plus errata sheet 1992), 23:13–27, which lays to rest earlier controversy about such cryptography.

8. Skelton, Marston, and Painter, *The Vinland Map;* D. B. Quinn and P. G. Foote, "The Vinland Map," *Saga-Book of the Viking Society* (London, 1966), 17:63–89; Samuel Eliot Morison, *The European Discovery of America: The Northern Voyages* (New York: Oxford Univ. Press, 1971), 69–72; Wilcomb E. Washburn, ed., *Proceedings of the Vinland Map Conference* (Chicago, 1971); Louis-Andre Vigneras, "Greenland, Vinland and the Yale Map," *Terrae Incognitae* (1972), 4:53–93; Leon Koczy, "The Vinland Map," *Antemurale* (1970), 24:85–171; Sandra J. Lamprecht, "The Vinland Map: A Selected Bibliography," *Special Libraries Association G & M Bulletin* (June 1988), 152:2–9; Enterline, "Cryptography."

9. Antoine de la Sale, "La Salade," *Bibliothèque de la Faculté de Philosophie et Lettres de l'Université de Liège* (Liège, 1935), 68:133–35.

10. Konrad Kretschmer, "Eine neue mittelalterliche Weltkart der vatikanischen Bibliothek," *Zeitschrift des Gesellschaft für Erdkunde zu Berlin* (Berlin, 1891), 26:371–406.

11. Durand, *Vienna-Klosterneuburg Map Corpus.*

12. Theobald Fischer, *Raccolta di Mappamundi e carte nautice del XIII al XVI secolo* (Venice, 1881), vol. 13; idem, *Sammlung Mittelalterlicher Welt- und Seekarten* (Venice, 1886), 213–38; Fridtjof Nansen, *In Northern Mists,* 2 vols. (London: Heinemann, 1911), 2:231–33.

13. George Kimble, *The Catalan World Map of the R. Biblioteca Estense at Modena* (London, 1934); *Monumenta Cartographica Vetustioris Aevi* (Amsterdam, 1964), pl. W and pp. 217–21; Konrad Kretschmer, "Die Katalanische Weltkarte der Biblioteca Estense zu Modena," *Zeitschriften der Gesellschaft für Erdkunde zu Berlin* (Berlin, 1897), 32:65–111, 191–218; F. Pulle and M. Loughend, "Illustrazione del Mappamondo Catalano del Bibliotece Estense di Modena," *Sèsto Congresso Geografico Italiano, Venezia 1907* (Venice, 1908).

14. Richard Henry Major, *The Voyages of the Venetian Brothers Nicolo and Antonio Zeno* (London, 1873), lxxv; *New York Times* (Aug. 25, 1968), 10; Arlington H. Mallery, *Lost America* (Columbus, Ohio, 1951), 163–68; Robert H. Fuson, *Legendary Islands of the Ocean Sea* (Sarasota, Fla.: Pineapple Press, 1995), 47–48.

15. Edward L. Stevenson, *Genoese World Map, 1457* (New York, 1912).

16. J. B. Harley and David Woodward, *The History of Cartography,* vol. 1 (Chicago: Univ. of Chicago Press, 1987), 312–13.

17. Tullia Gasparrini-Leporace, *Il Mappamondo di Fra Mauro* (Venice, 1956); Placido Zurla, *Il Mappamondo di Fra Mauro* (Venice, 1806); *Monumenta Cartographica Vetustioris Aevi,* 223–27.

CHAPTER 12 EUROPE'S WESTWARD AWAKENING

1. Helge Marcus Ingstad, *The Historical Background of the Norse Settlement Discovered in Newfoundland,* vol. 2 of *The Norse Discovery of America* (Oslo: Norwegian Univ. Press, 1985), 385–88; Sofus Larsen, *The Discovery of North America Twenty Years before Columbus* (Copenhagen, 1925), 29–37; David O. True, "Cabot Explorations in North America," *Imago Mundi* (1956), 13:25; Samuel Eliot Morison, *Portuguese Voyages to America in the Fifteenth Century* (Cambridge, Mass., 1940; New York, 1965), 37–41.

2. Miles Davidson, "The Toscanelli Letters: A Dubious Influence on Columbus," *Colonial Latin American Historical Review* (1996), 5 (3): 287–310.

3. David B. Quinn, *New American World* (New York, 1975), 1:82.

4. Samuel Eliot Morison, *Journals and Other Documents on the Life and Voyages of Christopher Columbus* (New York, 1963), 11–15; Thomas Goldstein, "Geography in Fifteenth-Century Florence," in *Merchants and Scholars* ed. John Parker (Minneapolis, 1965), 9–32; Thomas Goldstein, "Conceptual Patterns Underlying the Vinland Map," *Renaissance News* (1966) 19 (4): 321–31; John Thacher, *Christopher Columbus* (New York, 1903), 1:301–80.

5. A. E. Nordenskiöld, *Facsimile Atlas to the Early History of Cartography* (Stockholm, 1889; New York, 1964), 55–57, 86; idem, *Periplus* (Stockholm, 1897), 86–90; Joseph Fischer, *The Discoveries of the Norsemen in America, with Special Relation to Their Early Cartographic Representation* (London, 1903), 72–86; Joseph Fischer, "Claudius Clavus," *Historical Records and Studies, U.S. Catholic Historical Society* (New York, 1911), 74–85.

6. Vilhjalmur Stefansson, *Ultima Thule* (New York, 1940), chap. 2; Benjamin Keen, *The Life of the Admiral Christopher Columbus by His Son Ferdinand* (New Brunswick, N.J.: Rutgers Univ. Press, 1959), 11; J. K. Tornöe, *Columbus in the Arctic?* (Oslo, 1965; *Addendum,* 1967); A. B. Donworth, *Why Columbus Sailed* (New York, 1953); Alwyn A. Ruddock, "Columbus and Iceland: New Light on an Old Problem," *Geographical Journal* (London, 1970) 136:177–89.

7. Robert H. Fuson, *Legendary Islands of the Ocean Sea* (Sarasota, Fla.: Pineapple Press, 1995), 23–26.

8. Ibid., 72.

9. Davidson, "The Toscanelli Letters," 291.

10. Kirsten Seaver, *The Frozen Echo* (Stanford, Calif.: Stanford Univ. Press, 1996), chaps. 7, 8, and p. 225.

11. David B. Quinn, *England and the Discovery of America* (New York, 1974), chap. 13; James Williamson, *The Cabot Voyages and Bristol Discovery under Henry VII* (Cambridge, 1962), 19–23, 187–89, 211–14; James Howley, *The Beothuks or Red Indians* (Cambridge, 1915), 262; William Babcock, *Legendary Islands of the Atlantic,* (New York, 1922), chap. 4; E. Power, ed., *English Trade in the Fifteenth Century* (London, 1933), 161–81.

12. *Grönlands Historiske Mondesmaerker* (Copenhagen, 1845) 3:470.

13. Morison, *Portuguese Voyages to America,* 41–51; Charles Verlinden, "A Precursor of Columbus: The Fleming Ferdinand van Olmen (1487)," in *The Beginnings of Modern Colonization,* (Ithaca, N.Y., 1970), 181–95; Damiao Peres, *Descobrimentos Portugueses,* (Coimbra, 1960), 329–34.

14. Williamson, *The Cabot Voyages,* 23–24; Harry Kelsey, "The Planispheres of Sebastian Cabot and Sancho Gutiérrez," *Terrae Incognitae* (1987), 19:41–51; Seaver, *Frozen Echo,* 266–67, Appendix C.

15. Fischer, *Discoveries of Norsemen,* 49.

16. Seaver, *Frozen Echo,* 145 ff, 238.

17. E. G. Ravenstein, *Martin Behaim, His Life and His Globe* (London, 1908); G. R. Crone, "Martin Behaim, Navigator and Cosmographer: Figment of Imagination or Historical Personage?" *Congreso Internacional da Historia dos Descobrimentos, 1960: Actas* (Lisbon, 1961), 2:117–33; Heinrich Winter, "New Light on the Behaim Problem," *Congreso Internacional da Historia dos Descobrimentos, 1960: Actas* (Lisbon, 1961), 2:399–410; Wolfgang Pülhorn, ed., *Focus Behaim Globus* (Nuremberg, 1992).

CHAPTER 13 MASTERY OF THE ATLANTIC

1. Robert H. Fuson, *Legendary Islands of the Ocean Sea* (Sarasota, Fla.: Pineapple Press, 1995), 78, 101, 117.

2. John Noble Wilford, *The Mysterious History of Columbus* (New York: Knopf, 1991); Salvadore de Madariaga, *Christopher Columbus* (London, 1949); Samuel Eliot Morison, *Journals and Other Documents on the Life and Voyages of Christopher Columbus* (New York, 1963); Samuel Eliot Morison, *Admiral of the Ocean Sea* (Boston, 1942); John Thacher, *Christopher Columbus,* 3 vols. (New York, 1903–4).

3. E. G. Ravenstein, *Martin Behaim, His Life and His Globe* (London, 1908) 32–34, 113.

4. Franz R. von Wieser, "Die Karte des Bartolomeo Colombo . . . ," Mittheilung des Institut für österreichisch Geschichtsforschung, (Innsbruck, 1893), Supplement, 4:488–98; Kenneth Nebenzahl, *Atlas of Columbus and the Great Discoveries* (Chicago: Rand McNally, 1990), 38–39.

5. James Williamson, *The Cabot Voyages and Bristol Discovery under Henry VII* (Cambridge, 1962); Morison, *Journals and Other Documents;* Henry Harrisse, *The Discovery of North America* (London, 1892; Amsterdam: N. Israel, 1961); A. E. Nordenskiöld, *Facsimile Atlas to the Early History of Cartography* (Stockholm, 1889; New York, 1964); idem, *Periplus* (Stockholm, 1897); E. Lehner and J. Lehner, *How They Saw the New World* (New York, 1966); Edmundo O'Gorman, *The Invention of America* (Bloomington: Indiana Univ. Press, 1961); Germán

Arciniegas, *Amerigo and the New World* (New York, 1955); Robert H. Fuson, "The John Cabot Mystique," *Essays on the History of North American Discovery and Exploration* (College Station: Texas A&M Univ. Press, 1988), 35–51.

6. Ian Wilson, *The Columbus Myth* (London: Simon & Schuster, 1991), chaps. 9–11.

7. Wilcomb E. Washburn, "Representation of Unknown Lands in Fourteenth-, Fifteenth-, and Sixteenth-Century Cartography," *Agrupamento de Estudos de Cartografia Antiga* (Coimbra, 1969) 35:10, 16; Henry Stevens, *Historical and Geographical Notes* (New York, 1869; New York: Burt Franklin, 1970), 13, 14, 34.

8. Williamson, *The Cabot Voyages,* 72–83.

9. Afet Inan, *Life and Works of the Turkish Admiral Piri Reis* (Ankara, 1954); Gregory C. McIntosh, *The Piri Reis Map of 1513* (Athens: Univ. of Georgia Press, 2000), chap. 7.

10. Benjamin Keen, *The Life of the Admiral Christopher Columbus by His Son Ferdinand* (New Brunswick, N.J.: Rutgers Univ. Press, 1959), 25 and n. 3.

11. H. P. Biggar, *The Precursors of Jacques Cartier* (Ottawa, 1911), 31–99; Samuel Eliot Morison, *Portuguese Voyages to America in the Fifteenth Century* (Cambridge, Mass., 1940; New York, 1965), 51–72; Williamson, *The Cabot Voyages,* 116–40, 235–64; Arthur Davies, "Joao Fernandez and the Cabot Voyages," *Congreso Internacional de Historia dos Descobrimentos; Actas* (Lisbon, 1961), 2:135–49; R. A. Skelton, "The Cartographic Record of the Discovery of North America: Some Problems and Paradoxes," *Congreso Internacional de Historia dos Descobrimentos; Actas* (Lisbon, 1961), 2:243–363.

CHAPTER 14 A NEW CONTINENT EMERGES

1. Michael Layland, "The Line That Divided the World," *Mercator's World* (Eugene, Oreg., 1996), 1 (1): 34 ff.

2. James Williamson, *The Cabot Voyages and Bristol Discovery under Henry VII* (Cambridge, 1962), 116–40; 235–64; Arthur Davies, "Joao Fernandez and the Cabot Voyages," *Congreso Internacional de Historia dos Descobrimentos; Actas* (Lisbon, 1961), 2:135–49; R. A. Skelton, "The Cartographic Record of the Discovery of North America: Some Problems and Paradoxes," *Congreso Internacional de Historia dos Descobrimentos; Actas* (Lisbon, 1961), 2:343–63.

3. F. P. Sprent, *A Map of the World Designed by Giovanni Matteo Contarini* (London, 1926).

4. Wilcomb E. Washburn, "Representation of Unknown Lands in Fourteenth-, Fifteenth-, and Sixteenth-Century Cartography," *Agrupamento de Estudos de Cartografia Antiga* (Coimbra, 1969), 10.

5. G. B. G. Bull, review of *Maps of the World (Weltkarten) 1507 and 1517 by Waldseemüller,* ed. Joseph Fischer and R. von Wieser, *Geographical Journal* (1969), 135:144.

6. Joseph Fischer and Franz R. von Wieser, *The Oldest Map with the Name America of the Year 1507 and the Carta Marina of the Year 1516 by Martin Waldseemüller* (Innsbruck, 1903; Amsterdam, 1968).

7. C. Ptolemy, *Geographia,* (Rome, 1508); B. F. Swan, "The Ruysch Map of the World," *Bibliographical Society of America Papers* (1951), 45:219–36; J. Kenning, "Sixteenth-Century Cartography in the Netherlands," *Imago Mundi* (1952), 9:38; Donald L. McGuirk, Jr., "Ruysch World Map: Census and Commentary," *Imago Mundi,* 41:133–41.

8. H. P. Biggar, *The Precursors of Jacques Cartier* (Ottawa, 1911); Williamson, *The Cabot Voyages,* 145–72, 265–91; Eric Oxenstierna, *The Norsemen* (Greenwich, 1965), 96–122; Alwyn Ruddock, "The Reputation of Sebastian Cabot," *Bulletin of the Institute of Historical Research* (London, 1974), 47:95–99.

9. F. Kunstmann, *Die Entdeckung Amerikas* (Munich, 1859), no. 3; Axel A. Björnbo and

Carl S. Petersen, *Anecdota Cartographica Septentrionalia* (Havniae, 1908), no. 4; Heinrich Winter, "On the Real and Pseudo-Pilestrina Maps and Other Early Portuguese Maps in Munich," *Imago Mundi* (1947), 4:25.

10. C. Ptolemy, *Geographia,* (Strassburg, 1513).

CHAPTER 15 AN OLD CONTINENT EMERGES

1. M. Edme-François Jomard, *Les monuments de la géographie* (Paris, 1862), nos. 15, 16; Konrad Kretschmer, *Die Entdeckung Amerikas* (Berlin, 1891), pl. 13; Axel A. Björnbo and Carl S. Petersen, *Der Däne Claudius Clausson Swart* (Innsbruck, 1909), 52 ff.; Gustav Storm, "Den danske Geograf Claudius Calvus eller Nicolaus Niger," *Ymer,* (Stockholm, 1889), 9:138 ff.; Franz R. von Wieser, *Magalhaes-Strasse und Austral-Continent* (Innsbruck, 1881); Henry Harrisse, *The Discovery of North America* (London, 1892; Amsterdam: N. Israel, 1961), 491, 506–7; Edward L. Stevenson, *Terrestrial and Celestial Globes* (New Haven, 1921), 1:82–86; G. Sykes, "The Mythical Straits of Anian," *American Geographical Society Bulletin* (New York, 1915), 47:161–83.

2. James Enterline, "The Southern Continent and the False Strait of Magellan," *Imago Mundi* (Amsterdam, 1972), 26:48–58.

3. H. Stevens and C. Coote, *Johann Schöner* (London, 1888); Harrisse, *Discovery of North America,* 519, 528, 592–94; Lucien Gallois, *De Orontio Finaeo* (Paris, 1890); Stevenson, *Terrestrial and Celestial Globes,* 1:98, 107–14; Axel A. Björnbo, "Cartographia Groenlandica," *Meddelelser om Grönland* (Copenhagen, 1912), 48:243–50.

4. Jacob Ziegler, *Terrae Sanctae* (Strassburg, 1532); Herman Bror Richter, *Olaus Magnus's Carta Marina* (Uppsala, 1964; Lund, 1967); Edward Arber, *The First Three English Books on America* (Birmingham, England, 1885; New York, 1971), 295–303; Sofus Larsen, *The Discovery of North America Twenty Years before Columbus* (Copenhagen, 1925), 30–69; Sigmund Günther, *Jacob Ziegler* (Ansbach, 1896); Edward Lynam, *The Carta Marina of Olaus Magnus* (Jenkintown, Pa., 1949); Olaus Magnus, *A Compendious History of the Goths, Swedes and Vandals and other Northern Nations* (London, 1658).

CHAPTER 16 THE MISUNDERSTANDINGS ARE RESOLVED

1. Robert H. Fuson, *Legendary Islands of the Ocean Sea* (Sarasota, Fla.: Pineapple Press, 1995), 37–38; Donald S. Johnson, *Phantom Islands of the Atlantic* (Fredericton, N.B.: Goose Lane Editions, 1994), chap. 6.

2. Lucas, *Annals of Voyages,* 61.

3. Richard Henry Major, *Voyages of the Venetian Brothers Nicolo and Antonio Zeno* (London, 1873); Frederick W. Lucas, *Annals of the Voyages of the Brothers Zeno* (London, 1898); William Babcock, *Legendary Islands of the Atlantic* (New York, 1922), chap. 9; E. G. R. Taylor, "A Fourteenth-Century Riddle—and Its Solution," *Geographical Review* (New York, 1964), 54:573–76; Frederick Julius Pohl, "Prince 'Zichmini' of the Zeno Narrative," *Terrae Incognitae* (1970), 2:75–86.

4. Bert van't Hoff, ed., *Gerard Mercator's Map of the World (1569)* (Rotterdam, 1961); E. G. R. Taylor, "A Letter Dated 1577 from Mercator to John Dee," *Imago Mundi* (1956), 13:56–68; Lucas, *Annals of Voyages,* pl. 7; A. S. Osley, *Mercator* (London, 1969).

5. *Theatrum Orbis Terrarum* (Amsterdam, 1964), 1st ser., vol. 3 fol. 47; Cornelius Koeman, *The History of Abraham Ortelius and his Theatrum Orbis Terrarum* [with a reproduction of the

atlas] (New York, 1964); Arthur Middleton Reeves, *The Finding of Wineland the Good* (London, 1895; New York, 1967), 94–95.

6. Peter C. Hogg, "The Prototype of the Stefánsson and Resen Charts," *Historisk Tidsskrift* (Oslo: Universitetsverlag, 1989), 5.

7. Geoffrey M. Gathorne-Hardy, *The Norse Discoverers of America* (Oxford, 1921), 289–96; R. A. Skelton, Thomas E. Marston, and George D. Painter, *The Vinland Map and the Tartar Relation* (New Haven: Yale Univ. Press, 1965, 1995), 199; Hogg, "Prototype," 3–27.

CHAPTER 17 CONCLUSION

1. Benoit B. Mandelbrot, *The Fractal Geometry of Nature* (San Francisco: W. H. Freeman, 1982), 18.

2. Paul F. Hoyle and Paul Lunde, "Piri Reis and the Hapgood Hypothesis," *Aramco World Magazine* (1980), 31 (1): 18–31; Sean Mewhinney, "Minds in Ablation, Part 5," in MapHist. <MapHist@Harvarda.Harvard.edu>. July 15 and 16, 1998 (archived at <MapHist @Harvarda.Harvard.edu>. CD-ROM from: http://kartoserver.frw.ruu.nl/HTML/ STAFF/krogt/MapHist/cdrom.htm).

3. Persi Diaconis and Frederick Mosteller, "Methods of Studying Coincidences," *Journal of the American Statistical Association* (Washington, D.C.: American Statistical Association, Dec. 1989), 84:853–61.

4. Thomas S. Kuhn, *The Structure of Scientific Revolutions* (Chicago: Univ. of Chicago Press, 169).

5. James A. Bell, *Reconstructing Prehistory* (Philadelphia: Temple Univ. Press, 1994), 228–29.

6. David C. Hull, *Darwin and His Critics* (Chicago: Univ. of Chicago Press, 1983), 34.

7. Owen Gingrich, "Wonders," *Scientific American* (Sept. 1996), 183.

8. Robert A. Rundstrom, "Mapping, Postmodernism, Indigenous People, and the Changing Directions of North American Cartography," *Cartographica* (Toronto: Univ. of Toronto, 1991), 28:9.

9. Robert McGhee, "Disease and the Development of Inuit Culture?" *Current Anthropology* (1994), 35:565–94.

10. Knut Bergsland, "De norrøne lånord i grønlandsk," *Maal og Minne* (Oslo, 1986), 60.

11. Robert A. Hall, Jr., *The Kensington Runestone: Authentic and Important* (Lake Bluff, Ill.: Jupiter Press, 1994); Zalar, Michael, *The Kensington Runestone,* http://www.geocities.com/ m_zalar/index.html; Cyrus Gordon, *Riddles in History* (New York: Crown, 1974), 36–44, 119–44; Erik Wahlgren, "American Runes, from Kensington to Spirit Pond," *Journal of English and German Philology* (Champaign: Univ. of Illinois Press, 1982), 81 (2); Suzanne Carlson, "The Spirit Pond Inscription Stone: Rhyme and Reason," *New England Antiquities Research Association Journal* (1993–94), 28 (nos. 1–2): 1–7, (nos. 3–4): 74–82; Paul Chapman, "Spirit Pond Runestones: A Study in Linguistics," *Epigraphic Society Occasional Papers* (San Francisco, 1994), vol. 22, part 2.

12. Heather Pringle, "Hints of Frequent Pre-Columbian Contacts," *Science* (2000), 288:783.

13. R. A. Skelton, Thomas E. Marston, and George D. Painter, *The Vinland Map and the Tartar Relation* (New Haven: Yale Univ. Press, 1995), xxii–xxv.

14. Charles de La Roncière, *La carte de Christophe Colomb* (Paris: Édouard Champion, 1924); Monique Pelletier, "Peut-on encore affirmer que la Bibliothèque Nationale possède

la carte de Christophe Colomb?" *Revue de la Bibliothèque Nationale* (Paris, 1992), 45:22–25; Kenneth Nebenzahl, *Atlas of Columbus and the Great Discoveries* (Chicago: Rand McNally, 1990), 23.

15. Gunnar Thompson, *The Friar's Map of Ancient America* (Seattle: Argonauts of the Misty Isles & Laura Lee Productions, 1996), 104.

16. Harry Beilin and Peter B. Poufall, eds., *Piaget's Theory: Prospects and Possibilities* (Hillsdale, N.J.: L. Erlbaum, 1992), 31–33; Michael J. Mahoney, "Participatory Epistemology and Science," in *Psychology of Science,* ed. Shadish Gholson et al. (New York: Cambridge Univ. Press, 1989), 152–55.

17. Owen Gingrich, "Astronomy in the Age of Columbus," *Scientific American* (Nov. 1992), 100–105.

APPENDIX: THE VINLAND MAP'S INK

1. Helen Wallis et al., "The Strange Case of the Vinland Map," *Geographical Journal* (1974), 140:212–14.

2. Ibid., 212.

3. David Nunes Carvalho, *Forty Centuries of Ink* (New York: Banks Law Publishing, 1904), chap. 8, 163–64, 166–67, chap. 21; Wilhelm Wattenbach, *Das Schriftwesen im Mittelalter* (Leipzig: Hirzel, 1896), 239–40.

4. Wallis et al., "Strange Case," 213.

5. Ibid., 210, 214.

6. Wilcomb E. Washburn, ed., *Proceedings of the Vinland Map Conference* (Chicago: Univ. of Chicago Press, 1971), 35; Wallis et al., "Strange Case," 209–10.

7. Harold J. Plenderleith, *The Conservation of Prints, Drawings and Manuscripts* (Oxford, 1937), 5, 48.

8. W. F. Ehret, ed., *Smith's College Chemistry* (New York: Appleton-Century, 1946), 300, 304–5.

9. Plenderleith, *Conservation of Prints;* idem, *Conservation of Antiquities and Works of Art* (Oxford, 1956), 82.

10. Washburn, *Proceedings Vinland Map Conference,* 11.

11. W. R. Willets, "Application of Titanium Pigments to Paper and Board," *Paper Mill and Wood Pulp News* (Sept. 3, 1938); Kenneth Britt, ed., *Handbook of Pulp and Paper Technology* (New York: Reinhold, 1964), 415; "Paper Coating Pigments," *TAPPI Monograph Series,* no. 20 (New York: Technical Association of the Pulp and Paper Industry, 1958).

12. "Synthetic and Protein Adhesives for Paper Coating," *TAPPI Monograph Series,* no. 22 (New York: Technical Association of the Pulp and Paper Industry, 1961), 107 ff; Plenderleith, *Conservation of Prints,* 48.

13. Private correspondence to author from Lucy B. McCrone, Apr. 15, 1974.

SELECTED BIBLIOGRAPHY

Andrews, Michael C., "Study and Classification of Medieval Mappaemundi," *Archaeologia* (Oxford, 1926), 75:61–76.

Arctic Pilot, The, 3 vols. (London, 1915).

Bagrow, L., and Skelton, R. A., *History of Cartography* (Cambridge, 1964; Chicago: Precedent Publishing, 1985).

Beazley, C. R., *The Dawn of Modern Geography,* (Oxford, 1896–1906; New York, 1949).

Bekker-Nielsen, H., and Olsen, T. D., *Bibliography of Old Norse–Icelandic Studies, 1966* (Copenhagen, 1967).

Biggar, H. P., *The Precursors of Jacques Cartier* (Ottawa, 1911).

Björnbo, Axel A., "Cartographia Groenlandica," *Meddelelser om Grönland* (Copenhagen, 1912), vol. 48.

Brown, Lloyd, *The Story of Maps* (New York, 1949; New York: Dover Publications, 1979).

Congresso Internacional do Historia dos Descobrimentos, 1960: Actas (Lisbon, 1961).

Crone, G. R., *Maps and Their Makers* (London, 1953; Hamden, Conn.: Archon Books, 1978).

De Costa, B. F., "Arctic Exploration," *American Geographical Society Journal* (New York, 1880), 12:159–92.

Enterline, James R., *Viking America* (Garden City, N.Y.: Doubleday, 1972).

Fischer, Joseph, *The Discoveries of the Norsemen in America, with Special Relation to Their Early Cartographic Representation* (London, 1903).

Fischer, Theobald, *Sammlung Mittelalterlicher Welt- und Seekarten* (Venice, 1886).

Fossum, Andrews, *The Norse Discovery of America* (Minneapolis, 1918).

Fuson, Robert H., *Legendary Islands of the Ocean Sea* (Sarasota, Fla.: Pineapple Press, 1995).

Gade, John A., *The Hanseatic Control of Norwegian Commerce during the Late Middle Ages* (Leiden, 1951).

Geographical Journal, continuing series of Royal Geographical Society, London.

Geographical Review, continuing series of American Geographical Society; formerly A.G.S. *Bulletin* and A.G.S. *Journal.*

Globusfreund, Der (Vienna, annually since 1952).

Grönlands Historiske Mondesmaerker, 3 vols. (Copenhagen, 1845).

Hakluyt Society, *Works,* two series of several hundred volumes, continuing.

Harley, J. B., and Woodward, David, *The History of Cartography,* vol. 1 (Chicago: Univ. of Chicago Press, 1987).

Harrisse, Henry, *The Discovery of North America* (London, 1892; Amsterdam: N. Israel, 1961).

Harvey, P. D. A., *The History of Topographical Maps: Symbols, Pictures and Surveys* (London: Thames & Hudson, 1980).

Hennig, Richard, *Terrae Incognitae,* 4 vols. (Leiden, 1944–56).

Hoffman, Bernard G., *Cabot to Cartier* (Toronto, 1961).

Holand, Hjalmar R., *Explorations in America before Columbus* (New York, 1956).

Imago Mundi, continuing series, currently published in Amsterdam.

Ingstad, Helge, *Land under the Pole Star* (New York: St. Martin's, 1966).

Jones, Gwyn, *The Norse Atlantic Saga* (New York: Oxford Univ. Press, 1964).

Kimble, George H. T., *Geography in the Middle Ages* (London: Methuen, 1938).

———, *Geography of the Northlands* (New York: American Geographical Society and J. Wiley, 1955).

Kretschmer, Konrad, *Die Italienischen Portolane des Mittelalters* (Berlin, 1909).

Kuhn, Thomas S., *The Structure of Scientific Revolutions* (Chicago: Univ. of Chicago Press, 1962).

Mallery, Arlington H., *Lost America* (Columbus, Ohio, 1951).

Meddelelser om Grönland, continuing series with many articles in English (Copenhagen).

Morison, Samuel Eliot, *Admiral of the Ocean Sea* (Boston: Little, Brown, 1942).

———, *Journals and Other Documents on the Life and Voyages of Christopher Columbus* (New York, 1963).

———, *Portuguese Voyages to America in the Fifteenth Century* (Cambridge, Mass., 1940; New York, 1965).

———, *The European Discovery of America: The Northern Voyages* (New York: Oxford Univ. Press, 1971).

Muris, O., and Saarmann, G., *Der Globus im Wandel der Zeiten* (Berlin, 1961).

Nansen, Fridtjof, *In Northern Mists,* 2 vols. (London; Heinemann, 1911).

Oleson, Tryggvie, *Early Voyages* (Toronto; McClelland & Stewart, 1963).

Piaget, Jean, *Genetic Epistemology* (New York, 1970).

———, *Psychology and Epistemology* (New York, 1971).

Raemdonck, J. van, "Orbis Imago," *Cercle Archéologique du Pays de Waes, Annals* (1884), vol. 10.

Ramskou, Thorkild, *Vikingernes Hverdag / Everyday Viking Life* (Danish/English parallel texts) (Copenhagen, 1967).

Randles, W. G. L., *De la terre plate au globe terrestre* (Paris: Libraire Armand Colin, 1980).

Ristow, W., "Recent Facsimile Maps and Atlases," *Quarterly Journal of the Library of Congress* (Washington, D.C., 1967).

Ristow, W., and Le Gear, C., *A Guide to Historical Geography* (Washington, D.C., 1960).

Saga Book of the Viking Society for Northern Research, continuing series (London).

Schell, I. I., "The Ice off Iceland and the Climates during the Last 1200 Years, Approximately," *Geografiske Annaler* (Stockholm, 1961), 43:354–62.

Seaver, Kirsten A., *The Frozen Echo* (Stanford, Calif.: Stanford Univ. Press, 1996).

Simpson, Jacqueline, *Everyday Life in the Viking Age* (New York, 1967).

Skelton, R. A., *Maps: A Historical Survey of Their Study and Collecting* (Chicago: Univ. of Chicago Press, 1972).

Skelton, R. A., Marston, Thomas E., and Painter, George D., *The Vinland Map and the Tartar Relation* (New Haven: Yale University Press, 1965; republished with additions, New Haven, 1995).

Terrae Incognitae, continuing journal of the Society for the History of Discoveries (currently published in Arlington, Tex.).

Thomson, J. O., *History of Ancient Geography* (Cambridge, 1948; New York: Biblo & Tannen, 1965).

Thrower, Norman J. W., *Maps and Civilization: Cartography in Culture and Society* (Chicago: Univ. of Chicago Press, 1996).

Tooley, R. V., *Maps and Map Makers* (London, 1949, 1952).

Washburn, Wilcomb E., "Representation of Unknown Lands in Fourteenth-, Fifteenth-, and Sixteenth-Century Cartography," *Agrupamento de Estudos de Cartografia Antiga,* series 35 (Coimbra, 1969).

Woodward, David, "A Methodology for the Study of History of Cartography," *American Cartographer* (1975), vol. 1, no. 2.

Yonge, Ena L., "Facsimile Atlases and Related Material: A Summary Survey," *Geographical Review* (1963), 53:440–46.

FACSIMILE ATLASES AND REPRODUCTIONS

Alba, Duke of, (Jacobo Stuart Fitz-James of Falcó), ed. *Mapas españoles de América, siglos XV–XVII* (Madrid, 1951).

Almagia, R., *Monumenta Cartographica Vaticana* (Vatican City, 1944), vol. 1.

Bibliothèque Nationale, *Choix de Documents Géographiques* (Paris, 1883).

Björnbo, Axel A. and Petersen, Carl S., *Anecdota Cartographica Septentrionale* (Havniae, 1908).

Bratt, Einar, *En Kronika om Kartor över Sverige* (Stockholm, 1958).

Bricker, Charles, *Landmarks of Mapmaking* (Amsterdam, 1968; Oxford, 1981).

Caraci, Giuseppe, *Tabulae Geographicae Vetustioris in Italia Adservatae* (Florence, 1926–32).

Coote, Charles H., *Autotype Facsimiles of Three Mappemondes* (Aberdeen, 1896).

Cortesão, Armando, *Cartografia e cartografos portugueses* (Lisbon, 1935).

———, *Portugaliae Monumenta Cartographica* (Lisbon, 1960–62).

Destombes, Marcel, *Mappemondes: A.D. 1200–1500* (Amsterdam, 1964–)

Fischer, Theobald, *Raccolta di Mappamundi e carte nautice del XIII al XVI secolo,* ed. F. Ongania (Venice, 1881).

Fite, E., and Freeman, A., *A Book of Old Maps* (Cambridge, 1926).

Gasparini-Leporace, T., *Mostra: L'Asia nelle Cartografia degli Occidentali* (Venice, 1954).

Hantzsch, V., and Schmidt, L., *Kartographische Denkmäler zur Entdeckungsgeschichte* (Leipzig, 1903).

Harald Sigurdsson, *Kortsaga Islands* (Reykjavik, 1971, 1978, and 1981).

Henry E. Huntington Library and Art Gallery San Marino, California, *Collection of Photostat Reproductions of Manuscript Portolano Charts.*

Humphries, A. L., *Old Decorative Maps and Charts* (London, 1926).

Jomard, M., *Monuments de la Géographie* (Paris, 1854–55).

Kamal, Youssouf, *Monumenta Cartographica Africae et Aegypti* (Cairo, 1926–51; Frankfurt am Main, 1987).

Karpinski, Louis C., *Manuscript Maps Prior to 1800 Relating to America,* vols. 1, 2, 5, and indices.

Kretschmer, Konrad, *Die Entdeckung Amerikas* (atlas) (Berlin, 1892).

Kunstmann, Friedrich, *Atlas zur Entdeckungsgeschichte Amerikas* (Munich, 1859).

Landmarks of Early Cartography, continuing series of reproductions of entire atlases (Amsterdam, 1963–).

Lehner, E., and J. Lehner, *How They Saw the New World* (New York, 1966).

Lelewel, Joachim, *Géographie du Moyen Age* (atlas) (Brussels, 1852–57).

Marcel, Gabriel, *Reproductions de Cartes et de Globes* (Paris, 1893).

———, *Choix de Cartes et de Mappemondes* (Paris, 1896).

Miller, Konrad, *Mappaemundi: die ältesten Weltkarten* (Stuttgart, 1895–98).

———, *Mappae Arabicae* (Stuttgart, 1926–31).

Mollat, Michel, et al., *Sea Charts of the Early Explorers, Thirteenth to Seventeenth Century* (New York: Thames & Hudson, 1984).

Monumenta Cartographica Vetustioris Aevi (Amsterdam, 1964–).

Müller, Frederick, and Company, *Remarkable Maps* (Amsterdam, 1894–97).

Nebenzahl, Kenneth, *Atlas of Columbus and the Great Discoveries* (Chicago: Rand McNally, 1990).

Nordenskiöld, A. E., *Facsimile Atlas* (Stockholm, 1889; Vaduz, 1964; New York: Dover, 1973).

———, *Periplus* (Stockholm, 1897; New York: Burt Franklin, 1967).

Nörlund, N. E., *Danmarks Kortlaegning,* vol. 1 (Copenhagen: Ejnar Munksgaard, 1943).

———, *Islands Kortlaegning* (Copenhagen, 1944).

Rand McNally Company, series of maps and map reprints on Christmas cards (New York, at odd intervals).

Royal Geographical Society, *Reproductions of Early Manuscript Maps,* continuing series of full-sized color reproductions of individual maps (London, 1929–).

Santarem, Manuel, *Atlas* (Paris, 1842–53).

Sanz, Carlos, series of individual facsimiles (privately published at odd intervals).

———, *La Geographia de Ptolomeo* (Madrid, 1959).

———, *Mapas Antiquos del Mundo* (Madrid, 1962).

Shirley, Rodney W., *The Mapping of the World* (London: Holland Press, 1983, 1993).

Skelton, R. A., *Decorative Printed Maps* (London, 1952).

———, *Explorers' Maps* (New York, 1958).

Spekke, Arnolds, *The Baltic Sea in Ancient Maps* (Stockholm, 1961).

Stevens, Henry, *Historical and Geographical Notes on the Earliest Discoveries in America, 1453–1530* (London, 1869).

Suárez, Thomas, *Shedding the Veil: Mapping the Discovery of America and the World* (Singapore: World Scientific, 1992).

Theatrum Orbis Terrarum, several continuing series of reproductions of entire atlases (Amsterdam, 1963–).

Tooley, R. V., *Maps and Map Makers* (New York, 1949, 1952; London, 1970).

Wieder, F. C., *Monumenta Cartographica,* vol. 1 (The Hague, 1925).

World Encompassed, The. An Exhibition of the History of Maps, Held at the Baltimore Museum of Art, October 7 to November 23, 1952 (Baltimore, 1952).

INDEX

Reis, Piri, 23, 226
Renaissance
    definition, 10n
Resen, Hans, 286–290
Riphean mountains, 20, 58
Roes Welcome Sound, 44
Rolf the land settler ("Land-
    Rolf"), 113–114
Rossi, Marcian, 156
Ruin Island, 38
Rundstrom, Robert, 297
    Eskimo cartography, 14, 37
runestone
    Kensington, 297
    Kingigtorssuaq, 127–129
    Spirit Pond, 297
Russ, 20
Ruysch, Johann, 239–244
    bifurcated island as Brasil,
        68–69
    Inventio Fortunatae, 49

Sadlermiut, 45
Saenger, Paul, 63
sagas
    historicity, 101
    recorded, 101–102
Salmos, 140
Sanudo, Marino, 121–123
Satanaxes, 161
Saxo Grammaticus, 94
Scandinavia
    as island, 83, 121, 198
    as mainland, 94, 121, 198
    cartographic history, 16–20
Schöner, Johann. *See under*
    maps, descriptions,
    and voyages
Scolvus, Johannes, 195–196,
    247
Seaver, Kirsten, 132, 207, 211
    Vinland Map as forgery, 62
Septentrional Islands, 50, 52
    as Arctic Archipelago, 53–
        59
    on Behaim globe, 52
Seward Peninsula, 22
    in Clavus map, 35–40

Singui, 153
Skraelings
    definition, 21–22, 109n
    intense Norse interest in,
        109
    linguistically Karelian, 34
    reencountered in Green-
        land, 102–109
Snaefell, 103, 105
Söderberg, Sven, 64
solar compass, 77
solstice, winter, 21, 22
South America
    as Africa, 19, 93
Southampton Island
    in Clavus's second map,
        42–43
Spink, John
    Eskimo mapmaking, 12
Stefánsson, Sigurdur, 286–
        290
Strassburg Ptolemy, 250–251
structuralism, 7, 8, 17, 300
Stuttgart gores, 261
Sutherland, Patricia, 298
Svartenhavn, 105
Svartenhuk, 105
Szent-Gyorgy, Albert von, 4,
    6, 220, 299

Tabin peninsula, 264–265
Tabin Sea, 264–265
Taylor, E. G. R., 50, 67
terminology
    "accommodate to," 7
    "assimilate," 7
    "frigid Denmark"/"Ger-
        many," 121–122
    "Gateway to Greenland,"
        38
    "Grand Misunderstand-
        ing," 20
    "in the air," 7, 300
    "new historicism," 10
    "Pygmies" (for Eskimos),
        46
    "Smaller Misunderstand-
        ing," 20

"sun stone," 76
"tooth fairy rule," 24
"wishful seeing," 22
terra firma
    definition, 18
Thloyd (Lloyd?), John(?),
    206–208, 209
Thorfinn Karlsefni, 21, 87
Thorvald (priest), 113
Thule (place), 86, 88
Thule Eskimos, 22
    cartography, 12–15
    culture, 11, 11n, 35n
    definition, 11, 11n
    in *Gestae Arthuri,* 56
    migration, 11, 33, 38, 102
titanium
    presence on Vinland Map,
        305–309
Tobler, Waldo
    image-comparison tech-
        niques, 40–41
Toscanelli, Paolo, 196–198,
    206

Unartoq hot spring, 143
Ungava Bay, 264–265
Uvkusigssat, 105

Vesconte, Petrus, 121
Vespuccius, Americus, 220
Victoria Island, 56
Vienna Klosterneuburg New
    Cosmography, 169–
        170
Vienna Klosterneuburg
    Septem Climatum,
        167–169
Vienna Klosterneuburg
    Schyfkarte, 183–185
Vienna text (of Clavus), 41–
    42, 45–46
Vinilanda Insula, 61, 64, 65–
    67, 68, 70
Vinland, 87, 87n
    and Africa, 93
    as pastureland, 64–65, 90–
        91